Synthesis Lectures on Engineering, Science, and Technology

The focus of this series is general topics, and applications about, and for, engineers and scientists on a wide array of applications, methods and advances. Most titles cover subjects such as professional development, education, and study skills, as well as basic introductory undergraduate material and other topics appropriate for a broader and less technical audience.

Md Rafiul Kabir · Sandip Ray

Digital Twins for Distributed IoT Applications

Springer

Md Rafiul Kabir
Central Michigan University
Mount Pleasant, MI, USA

Sandip Ray
University of Florida
Gainesville, FL, USA

ISSN 2690-0300 ISSN 2690-0327 (electronic)
Synthesis Lectures on Engineering, Science, and Technology
ISBN 978-3-032-18937-0 ISBN 978-3-032-18938-7 (eBook)
https://doi.org/10.1007/978-3-032-18938-7

This Springer imprint is published by the registered company Springer Nature Switzerland AG
The registered company address is: Gewerbestrasse 11, 6330 Cham, Switzerland

To

Md Khairul Kabir and Nahid Akhter
Anindita Das and Aapti Ray
Who showed us that dreams can come true

Preface

The rapid growth of the Internet of Things (IoT) has fundamentally transformed the design, monitoring, and optimization of physical systems. Across domains as diverse as energy, transportation, healthcare, and manufacturing, IoT has enabled unprecedented connectivity between cyber and physical entities, giving rise to increasingly sophisticated, data-rich, and distributed cyber-physical systems. Meanwhile, traditional modeling and simulation tools, which have long been used to examine such systems, are proving unable to meet the scale, dynamism, and real-time demands of modern IoT ecosystems. It is within this evolving technological landscape that the idea of the *digital twin* has emerged as a strong and unifying tool.

This book was conceived to address a growing need for a comprehensive, system-level perspective on digital twins and related virtual prototyping technologies, particularly as they apply to distributed IoT environments. While digital twins have gained significant attention in recent years, the literature is still dispersed and limited to specific domains, narrowly focused use cases, or isolated technical viewpoints. It can be difficult for academics, practitioners, and students to comprehend how digital twins relate to more general prototyping concepts, how they are enabled by IoT infrastructures, and how they can be systematically applied across diverse application domains. When a student or a new researcher asks, "So what exactly counts as a digital twin here?" the honest answer is: it depends on context. That dependency is precisely what we wanted to clarify.

The primary purpose of this book is, therefore, not to present digital twins as a buzzword or a single blueprint, but to place them inside a broader prototyping ecosystem and explain how they behave in *distributed IoT settings*. We emphasize that the role of sensing, connectivity, data analytics, and cyber-physical integration collectively enable digital twins to evolve from static virtual models into dynamic, adaptive, and intelligent counterparts. Throughout the book, we focus not only on what digital twins are, but more importantly, how they are used, how they differ from closely related technologies, and what practical challenges and opportunities they introduce.

The scope of the book is structured but intentionally broad. It begins by establishing foundational concepts in virtual prototyping and digital twin technologies, clarifying terminology that is often used inconsistently across academic and industrial contexts. Building on this foundation, the book explores the role of digital twins in several major IoT-driven domains that are shaping modern society. These include smart energy systems, intelligent vehicular systems across automotive, aviation, and maritime sectors, smart healthcare and the Internet-of-Medical Things (IoMT), and IoT-enabled smart manufacturing under the Industry 4.0 paradigm. Each domain is explored through a narrow lens, emphasizing domain-specific requirements, application patterns, and unresolved research challenges.

The book is organized into five chapters, each designed to be relatively self-contained while contributing to a cohesive overall narrative. The first chapter provides a comprehensive overview of virtual prototyping technologies for distributed IoT systems, positioning digital twins alongside related approaches such as virtual platforms and application-specific virtualization frameworks. The following chapters dive into domain-centric perspectives, exploring how digital twins are used in energy systems, vehicular systems, healthcare IoT, and smart manufacturing. This structure allows readers to either progress sequentially or focus on specific chapters relevant to their interests.

The intended audience for this book is broad and multidisciplinary. It is written for a mixed audience: researchers who want a consolidated perspective, graduate students who want clarity and research directions, and practitioners who want to understand how these ideas map to production-grade real systems. While the book emphasizes conceptual frameworks, taxonomies, and system-level insights, it deliberately avoids unnecessary mathematical and implementation-specific detail, making it accessible to readers from diverse technical backgrounds. It is intentionally not a "tool manual." Instead, it aims to give readers a reliable mental model: what to build, what to watch for, and what questions to ask when designing or evaluating digital twins for distributed IoT applications.

This book is the result of years of research, teaching, and collaboration across multiple institutions, projects, and disciplines. We are deeply grateful to our collaborators, colleagues, and students whose discussions, critiques, and contributions helped shape many of the concepts presented here. We also acknowledge the broader research community whose prior work served as the basis for this book. Finally, we extend our sincere appreciation to our family for their patience and support during the writing process.

We hope that this book will serve not only as a reference on the current state of digital twins for distributed IoT applications but also as a driver for future research, innovation, and cross-disciplinary collaboration in this rapidly evolving field. Digital twins are still evolving, and we're excited to see where they go next.

Mount Pleasant, MI, USA
Gainesville, FL, USA
December 2025

Md Rafiul Kabir
Sandip Ray

Acknowledgements The authors would like to sincerely acknowledge Dipal Halder, Fairuz Shadmani Shishir, Sumaiya Shomaji, and Bhagawat Baanav Yedla Ravi for their valuable collaboration on several survey papers that informed and influenced parts of this book. Their contributions, discussions, and joint research efforts have helped shape the ideas and technical perspectives presented here.

Competing Interests The authors have no conflicts of interest to declare that are relevant to the content of this chapter.

Contents

Acronyms

ADAS	Advanced Driver Assistance Systems
AR	Augmented Reality
ATE	Automated Test Equipment
AV	Autonomous Vehicle
B5G	Beyond 5th Generation
BEM	Building Energy Modeling
BESS	Battery Energy Storage System
BIM	Building Information Modeling
BMS	Battery Management System
CAD	Computer-Aided Design
CAE	Computer-Aided Engineering
CAM	Computer-Aided Manufacturing
CCA	Coordinated Cyber Attack
CEMS	Cloud-based Energy Management System
CloudPSS	Cloud Power System Simulator
CNN	Convolutional Neural Network
CPES	Cyber-Physical Energy Systems
CPPS	Cyber-Physical Power Systems
CPS	Cyber-Physical System
DCU	Domain Controller Unit
DER	Distributed Energy Resources
DR	Demand Response
DT	Digital Twin
ECU	Electronic Control Unit
EDT	Energy Digital Twin
EMA	European Medicines Agency
ERPS	Electric Railway Power System
EV	Electric Vehicle

FDA	Food and Drug Administration
FHIR	Fast Healthcare Interoperability Resources
GAN	Generative Adversarial Network
GDPR	General Data Protection Regulation
GDTA	Generic Digital Twin Architecture
HDT	Human Digital Twin
HIL	Hardware-in-the-Loop
HIPAA	Health Insurance Portability and Accountability Act
HL7	Health Level Seven
ICD	Implantable Cardioverter Defibrillator
ICU	Intensive Care Unit
IIoT	Industrial Internet of Things
IMD	Implantable Medical Device
IoMT	Internet of Medical Things
IoT	Internet of Things
IoV	Internet of Vehicles
LLM	Large Language Model
LSTM	Long Short-Term Memory
ML	Machine Learning
OADT	Online Analysis Digital Twin
OME	Observable Manufacturing Elements
PBTES	Packed-Bed Thermal Energy Storage
PHEVs	Plug-In Hybrid Electric Vehicles
PLM	Product Lifecycle Management
PSDT	Power System Digital Twins
PV	Photovoltaics
RAMI	Reference Architecture Model for Industry 4.0
RESS	Residential Energy Storage Systems
RMES	Regional Multiple Energy System
ROS	Robot Operating System
RTDT	Real-Time Digital Twin
SIL	Software-in-the-Loop
SoC	System-on-Chip
SOC	State of Charge
SOH	State of Health
UAV	Unmanned Aerial Vehicle
V2G	Vehicle-to-Grid
VP	Virtual Platform
VR	Virtual Reality
WMSN	Wireless Multimedia Sensor Network

1 Introduction to Prototyping: Definitions, Concepts and Enabling Technologies

Digital Twins as a Foundation for Modeling and Exploring Cyber-Physical IoT Systems.

1.1 Introduction

The Internet-of-Things (IoT) has gained popularity over the past ten years by evoking the idea of a worldwide infrastructure of networked physical things that would provide anytime, anywhere connectivity for anything and not just for any one person [1]. As technology evolves, more and more cyber components are getting integrated into the physical systems around us, incorporating more aspects of IoT. Embedded systems are used in most applications that monitor all physical mechanisms affecting computations and *vice versa*. Naturally, the integration of cyber and physical components of systems came up with terms and explored new vision i.e., in the modern-day, what we call "a cyber-physical system" (CPS) [2–4]. Furthermore, IoT revolves around "smart systems"—an ability to gather and apply knowledge independently—referring to the "things and sensors" that are intelligent, uniquely addressable, flexible, and autonomous with intrinsic security [5]. A shortened term for the industrial applications of IoT, commonly referred to as the Industrial Internet-of-Things (IIoT), is another concept reshaping the modularly structured smart factories of Industry 4.0 [6]. Additionally, conventional domains such as control systems, wireless sensor networks, and automation enable IoT systems both individually and collectively, particularly in the areas of privacy and security. Combining these concepts, IoT is revolutionizing modern industrial applications into a smarter domain, and more emerging technologies are being developed to contribute to that expansion.

The system-level exploration of IoT systems has gained significant attention nowadays. The spread of new technologies connected through the Internet inspired researchers to study

M. R. Kabir and S. Ray, *Digital Twins for Distributed IoT Applications*, Synthesis Lectures on Engineering, Science, and Technology, https://doi.org/10.1007/978-3-032-18938-7_1

physical systems at an abstract level. While modeling and simulation have been around with the rapid growth of the computer for several decades, structured prototyping of systems, sub-systems, and components as a virtual counterpart is a recent development [7]. There are several terms in common industrial use—*virtual prototype, virtualization, virtual platform, digital twin, virtual replica, digitalization, modeling, simulation, etc.* that, albeit similar, have somewhat nuanced and different interpretations and apply to slightly different contexts. In the context of IoT and software modeling, digital twin is an engineering discipline that involves modeling, simulating, visualizing, or abstracting a software, hardware, or overall physical system based on its fully functional operating behavior [8]. It allows any real-world system or component to be mimicked into a virtualized version that can be explored freely. The integration of instruction-set simulators, hardware building block models, numerical analysis, and twining techniques has motivated us to achieve more advancement.

In spite of its critical need, only a limited number of comprehensive books currently provide a generic overview of different kinds of virtual prototyping technology, including digital twin as the front-runner for various IoT applications in the industry. Consequently, it is a daunting task for a researcher or practitioner getting started in this area to sift through the mass of research articles across journals and conference proceedings, often spanning disparate applications with their own rich bodies of literature, to assimilate the information about the state of the art. In this book, our goal is to fill this crucial gap. We explore the variety of prototyping techniques, including digital twins, virtual platforms, and various domain-specific infrastructures, point out the key differences and the contexts of their application.

1.2 Technical Challenges

1.2.1 Exploration of Cyber-Physical Systems

The design of any cyber-physical system is a challenging task as it requires going over a set of CPS configurations involving software, hardware, and integration. Major CPS, like autonomous automotive systems, medical monitoring, smart grids, industrial automation, etc., face hard-to-detect errors that can induce major drawbacks later in the design, compromising reliability, efficiency, and safety. This demonstrates the inherent complexity of dealing with the design of a CPS, as the types and heterogeneity of the components can vary across domains and applications [9]. With current industrial practice, it is hard to resolve design errors and bug fixes before the assembly and manufacturing of hardware and software components. The need for virtual prototyping emerges to solve this critical problem by exploring the corresponding cyber-physical system way early in the system life-cycle, which can eliminate potential design errors. The CPS involves heterogeneous components that require close interactions in a virtual environment to determine several exploration aspects. The developed prototyping techniques will be readily available to the design engineers of embedded control systems to help them explore and better understand the overall system for both hardware and software functionalities.

The majority of current CPS development focuses on the physical layer of embedded systems or the potential applications of the CPS domain [10]. Without a clear bridge, it is unclear how the embedded systems of the physical layers will be used to supply real-time services to the application layer. The virtual prototypes of such systems aim to eliminate these discrepancies by establishing real-time monitoring and diagnostics. The prototype computation modules analyze associated data, inform the physical systems of their results, and if necessary, transmit control commands to modify the physical environment or adjust system parameters [11, 12].

1.2.2 Optimization

When considering an IoT-based application as a dynamic operating system, optimization is essential for improving reliability, efficiency, and the application of key process parameters, thus ensuring a better operation. The option of prototyping provides the appropriate framework to adaptively tune certain parameters without meddling with the actual system design i.e., allowing us to perform seamless optimization techniques. Moreover, the integration of virtual prototyping and optimization approaches can obviously provide a promising tool for quickly resolving errors and making informed decisions. The best approach is to build a prototype that supports real-time synchronization of IoT components, cyber backbone, and physical systems. For instance, different data analytics can be integrated to manage data on the machine by establishing digital twins of the machining process (i.e., mostly studied in manufacturing systems) [13].

Optimizing the physical one based on the virtual model is a well-practiced approach, especially in modern IoT industries like aviation (e.g., tires, aircraft health, etc.) and manufacturing (e.g., shop floor, plant automation, etc.). Prototyping frameworks can also help to improve the robustness of the cyber-physical system by providing the interdependence of cyber and physical networks to explore various nonlinear optimization models [14]. There have been studies that focused on the optimization of various domains like—virtual reality [15], simulation of virtual platforms [16, 17], network function virtualization [18, 19], digital twin for production [20–22], etc. These approaches can eventually result in higher efficiency, highly advanced, and intelligent IoT systems with less complexity, provided the optimization techniques perform well.

1.2.3 Safety and Security

Current industry practice incorporates only a partial form of safety and security based on isolated networks and access control environments [23]. Cybersecurity is a major concern in IoT-enabled systems, which may be vulnerable to a wide range of cyberattacks from potential adversaries. As a result, cybersecurity is essential for smart systems to succeed in various industry applications. Theft of trade secrets and intellectual property, hostile data modifications, and disruptions or denial of process control are all genuine, worldwide, and growing cyber risks to the Industrial IoT systems [24]. Usually, security assumptions are made to design and operate the IoT infrastructure that might be at risk of not being fully technologically capable of eliminating potential security threats. Given this insight, virtual prototyping is implemented to evaluate the security vulnerabilities of primary system design, security software, and other IoT configurations, aiming to make them secure against first-order attacks [25]. A prototyping infrastructure can help to explore situations of accidental failure in order to avoid hazards and thus maintain the overall system's safety.

Several studies have attempted to address the safety and security needs using prototyping of respective IoT systems. For instance, Almeaibed et al. [26] sought to find a common framework for digital twins of vehicles that facilitates the analysis of a vehicle follower model to promote safety and security in autonomous vehicles. Alcaraz et al. [27] explored the current state of the DT paradigm by categorizing potential threats associated with it and offering a preliminary set of security recommendations. Lou et al. [28] proposed conducting a functionality and cybersecurity analysis based on the digital twin of an Industrial Control System (ICS). We understand that, when aligned with a virtual framework, the IoT infrastructure can be explored independently and function together to provide a solid foundation for an intelligent IoT application, while eliminating any shortcomings in the safety and security assessment.

1.3 Virtual Platforms

Before introducing system-level digital representations for IoT and cyber-physical systems, it is useful to first examine the concept of Virtual Platforms (VPs), which have a long history in embedded systems and hardware-software co-design. VPs represent one of the earliest and most mature forms of virtual system modeling, particularly at the component and subsystem level. Understanding their scope, capabilities, and limitations provides an important foundation for later discussions on more comprehensive digital representations used in distributed and data-driven systems. It is a software-based modeling system that can completely mimic the functionality of a specific SoC or board. It can combine high-speed processor simulators with high-level, fully functional models of hardware building blocks in order to provide software developers and system architects with an abstract, reconfigurable representation of the hardware. This approach has been valuable for early software and hardware development, verification and validation, and hardware-software co-design [29, 30]. Because of their high levels of controllability and observability, VPs today provide a powerful debug infrastructure [31]. While VPs are used mostly for the smaller sub-systems (e.g., SoCs), they indirectly contribute to the larger IoT infrastructures that incorporate the target components being used for developing VPs.

1.3.1 Virtual Platforms for SoC and Embedded Software

Modern cyber-physical systems incorporate embedded electronic components into a single integrated circuit that is known as a system-on-chip (SoC). The trend of SoC usage has driven constant technological advancements as it contains multiple processors combined within dedicated hardware. For rapid SoC development at an abstraction level, the need for a configurable virtual platform emerged. Today, the VPs are constructed using SystemC and transaction-level modeling (TLM) to further increase the abstraction level [32]. Given this insight, the authors conceptualized a virtual platform, *SoCRocket* [16], for the rapid SoC development in the aerospace domain. They developed a TLM-based framework where the core components have been modeled in SystemC. Their platform is available in three abstraction levels i.e., loosely timed (LT), approximately timed (AT), and register-transfer-level (RTL). An integrated framework *SPHERE* [33] is introduced, targeting modern SoCs by abstracting the hardware complexity and virtualizing computational resources. SPHERE aims for the smooth operation of different subsystems on the same platform while providing a safe and secure functioning of cyber-physical system mechanisms.

VPs allow for early exploration and optimization of the embedded software development process by providing a high-speed framework i.e., the ability to perform fast system simulations with only necessary functionalities in real-time. This means that the VP simulation framework must be fully instruction-accurate, with all interrupts and similar commands correctly emulated, and the peripherals and behavioral models must provide the proper functionality. Hong et al. [34] presented a case study of the VP application for a new hard disk system development known as *Hybrid-HDD* i.e., one of the primary features of Windows VISTA. They summarized their model by comparing it with the conventional flow of software development, and explored software optimization via the VP. A conventional virtual platform activity in terms of three domains is depicted in Fig. 1.1 portraying the enabling technology of prototyping for mixed hardware/software development.

1.3.2 Virtual Platforms for Hardware/Software Co-Design

Throughout the SoC and embedded system design flow, verification and validation procedures ensure the requirements and specifications of software quality management and hardware/software co-design. Nowadays, researchers are moving towards using virtual prototyping platforms to solve both co-design and co-verification problems. A conventional VP for validation and hardware/software co-design is shown in Fig. 1.2, where the basic actions of the prototyping environment are depicted as a flow diagram. The modeling and simulation come into effect with the reference information and HW-SW configuration data for prototyping. After that, the virtual output gets validated based on the specifications of the prototyping basis and moves towards optimization. The invalid model goes through the HW/SW co-design process for refining, to be used again for simulation.

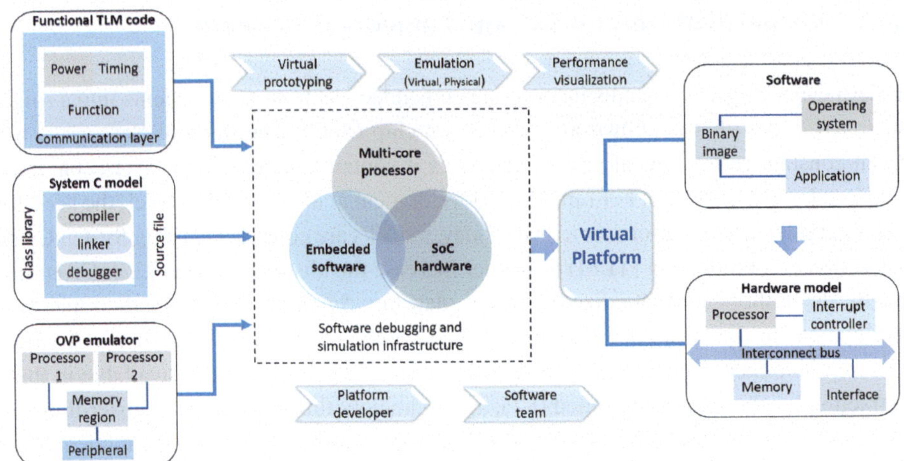

Fig. 1.1 A conventional virtual platform activity. The functional TLM, SystemC models, and OVP emulators are the most common modes of developing platforms. The areas that utilize this technology are multi-core processors, embedded software, and SoC hardware. Abstraction of software runs on top of a hardware model that simulates the designated virtual platform

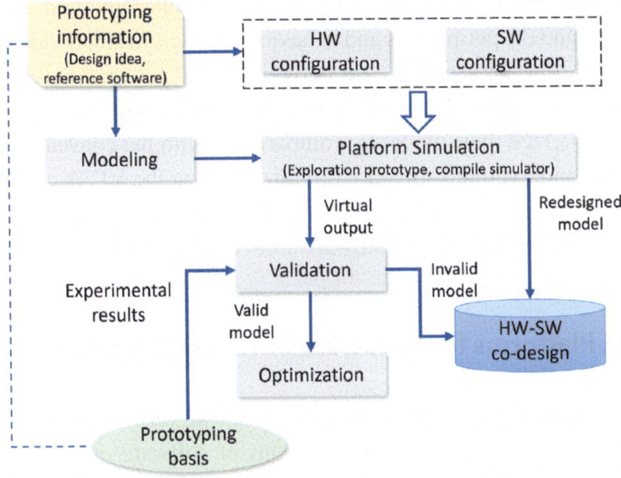

Fig. 1.2 VP for validation and HW-SW co-design

Lin et al. [35] discussed a heterogeneous virtual simulation platform targeting the system and functional level co-verification of SoC Software/Hardware Co-Design. They focused on functional and system-level verification by using an open-source emulator, *QEMU* to perform co-simulation within the SoC design flow. There are several researchers [36–40], who established virtual platforms using *QEMU* due to its ability to provide high-performance CPU emulation. Wicaksana et al. [41], developed a virtual platform using SystemC with TLM and the Open Virtual Platforms (OVP) processor model with an instruction set simulator (ISS) targeting a multiprocessor system-on-chip (MPSoC) to fulfill the hardware/software co-design and verification requirements. Analogous to the conventional approach, they increased the abstraction level of the SoC design and verification to the electronic system level (ESL).

1.3.3 Virtual Platforms for Cyber-Physical Configuration and Networks

Other than focusing on SoCs and embedded applications, there have been a few virtual platforms that facilitate explorations of the overall system framework and network architecture. Ahn et al. [42] introduced a virtual platform named *Xebra*, targeting the need for a global-scale cyber-physical system by virtualizing the framework and isolation techniques that include CPS middleware. Their platform is unique in the sense that it allows the coexistence of various global-scale IoT solutions on a physical network through virtualization. To support the network among various CPS applications, they implemented low-overhead message control by managing a layered virtual network. The network virtualization and its corresponding mapping to the system architecture are depicted in Fig. 1.3.

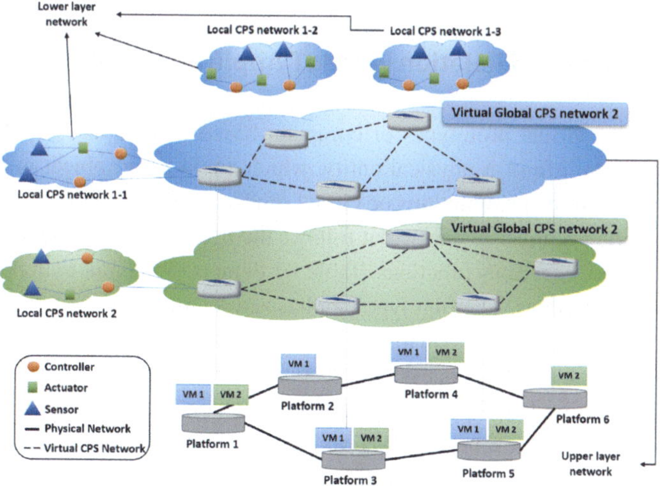

Fig. 1.3 Network virtualization and its corresponding mapping to the system architecture [42]

Soares et al. [43] introduced a virtual platform for cloud-based network virtualization that is called Cloud4NFV. The platform is focused on Virtual Network Functions (VNFs) that follow the network architecture guidelines and aim to deliver a new service to end customers, emphasizing customer premises equipment (CPE) related functions. Furthermore, some other works have focused on the virtualization of specific network features, e.g., Network Functions Virtualization (NFV) enhanced with Software Defined Networking (SDN) [44] and virtualized routing functions [45]. Necessary network functions like load balancing, routing, and firewall security are all performed by an NFV-based virtual framework instead of the hardware components. These studies have demonstrated that the potential benefits of NFV are expected to be substantial. Virtualization of network functions on general-purpose standardized hardware is projected to minimize capital and operational costs, as well as the time it takes to launch new services and products.

1.4 Digital Twins

The concept of *digital twin* (DT) [46] has gained significant popularity for building advanced cyber-physical systems, smart systems, and IoT-infused applications. The idea was conceptualized by Grieves [47] as the conceptual paradigm underlying product lifecycle management. Unlike virtual platforms, digital twins focus on the physical behavior of the physical entity rather than abstracting any software layer. They offer several benefits, like simulation, prediction, and monitoring, once they combine the physical and virtual assets through the Internet-of-Things. Many important industries, including the manufacturing sector, connected and autonomous vehicles, healthcare, energy, city planning, and many more, are being revolutionized by DT technologies [48].

1.4.1 Definitions and Relation with IoT Applications

DTs are defined as virtual prototypes or computer-based models that simulate, emulate, mirror, or "twin" the life of a physical entity, which could be an object or process [49]. DTs are not limited to just simulations or virtual models [50]. It is an intelligent, evolving digital counterpart of a physical entity that follows the life cycle of its physical counterpart to monitor and process various functions. Table 1.1 depicts the digital twin definition from three different perspectives. The first type emphasized the mirroring or virtual representation of a physical object or process. This definition, however, ignores any automated data or simulation perspective on that process. The second definition type focused mainly on portraying the DT as a simulation process, automated data, or a prediction model. According to this definition, the data flow between DT and the physical entity is unidirectional i.e., any change in the physical object will affect the virtual one, but not the other way around [51].

Table 1.1 Digital twin definitions

Domain	Definition	Refs.
Digital counterpart, Model, Twin	The virtual and computerized counterpart of a physical system	[52]
	DT is a virtual representation of a physical product or process, used to understand and predict its physical counterpart's performance characteristics	[53]
	A dynamic digital representation of a physical system	[54]
Simulation, prediction	A simulation based on expert knowledge and real data collected from the existing system	[55]
	Digital twin is to develop the virtual models for physical objects in order to simulate their behaviors	[56]
	Re-engineering computational model of structural life prediction and management	[57]
Integration	Comprehensive physical and functional description of a component, product, or system together with available operational data	[58]
	Integrated multi-physics, multi-scale, and probabilistic simulation compost or physical products, virtual products, data, services, and connections between them	[59, 60]

DT is defined as an integration process in the third type of definition. Here, DT incorporates both the physical object and its corresponding virtual model and continually adapts based on the connection between these two.

For the improvement of the IoT system in every aspect, DT can be constructed from any physical entity that can provide feedback based on simulation results. As discussed in Sect. 1.2, virtual prototyping i.e., DT for this discussion, can help develop the IoT application in terms of exploration, optimization, and security aspects. Figure 1.4 depicts the construction of DT from a typical industrial IoT that is showing feedback for adaptation, and also different kinds of DT models that can be involved. A typical industrial IoT system contains an IoT-based process, data storage, and decision-making system integrated with DT information flow. Here, DT computation modules process the virtual data after proper visualization and simulation, along with human inputs if necessary, and notify back to the physical system about the findings from the simulation to make necessary changes in the physical version or adapt system parameters if necessary [12].

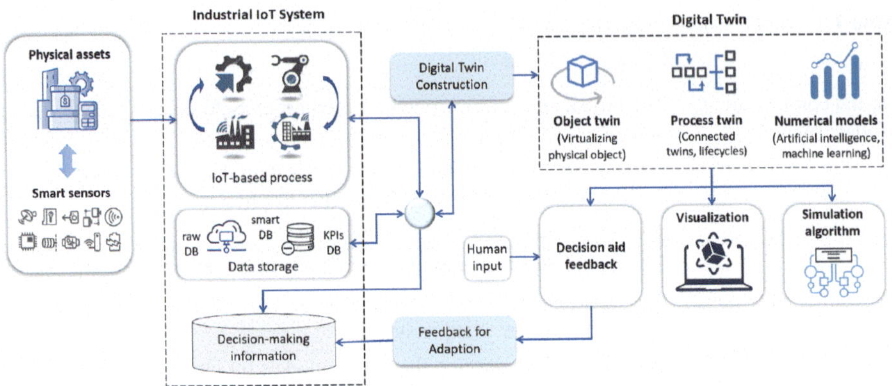

Fig. 1.4 Digital twin construction for industrial IoT systems

1.4.2 AI and ML-Based Numerical Models

Smart data analysis for IoT-based systems using artificial intelligence (AI) and machine learning (ML) has great potential in the context of prototyping with the digital twin. We mentioned the DT definition as an integrated system; the numerical model-based frameworks contribute to this purpose. Smart systems include various IoT devices that expedite real-time data processing i.e., leading to the development of digital twins integrated with AI-ML techniques. As a result, methods like Generative Adversarial Networks (GANs) [61] and Restricted Boltzmann Machines (RBM) [62] are having a great impact on improving data quality and understanding time series with digital twin [63]. In some industries, the DT framework combines IoT, data analysis, and machine learning for prediction, monitoring, correction, and comparison activities to improve and control the existing IoT components. While developing complex cyber-physical systems, trial versions seem too expensive to deploy, especially in developing countries. In order to accurately predict the outcome of any system in advance, specific simulators are often used that are designed to replicate the physical system. These simulators are usually developed based on independent computational models i.e. artificial intelligence concepts that require self-configuring data. This is where the DT can be of great potential, as the goal of a virtual model within a DT is to self-learn the data pattern in order to optimize the configuration of a physical system in real-time [64].

DT technology refers to the supervised and unsupervised learning algorithms that refine their predictive ability as they process continuously acquired sensed data from the physical twin and the surrounding environment [49]. Here, the virtual twin acts as an intelligent representation of the cognitive brain that performs a set of tasks via predictive algorithms. Feature selection and feature extraction methods play an important role in big data by reducing the data range and extracting only the informative data i.e., enabling effective real-time system synchronization. Key AI-ML algorithms like pattern recognition, unsupervised and supervised learning, and statistical applications let the DT characterize, analyze, cluster, and classify input data from the surrounding CPS environment [65]. Analyzing data enables the detection of changes as well as the identification of relevant patterns and trends. Thus, various businesses are setting up new efficiencies as AI-powered digital agents can predict impending asset failure and the underlying causes weeks in advance. Industrial organizations can then use digital twins to propose process actions to maintain equipment health and reduce plant downtime.

Droder et al. [66] proposed a machine learning-based digital twin for human-robot collaboration to eliminate the problem of safe movement of the robot in an unstructured human environment. To develop and test their approach, they developed a digital twin using MATLAB that is also extended by integrating with artificial neural networks (ANNs) [67]. A production system-based IoT framework for a digital twin combining machine learning and simulation is presented in another paper [68]. They used finite element method (FEM) simulation for the digital twin and adapted it with machine learning-based surrogate modeling of the FEM. Zhou et al. [69] presented an AI-based detection model for small objects via digital twin, aiming to dynamically synchronize a physical system with its virtual representation. For a smart manufacturing framework, they built their DT based on three parameters i.e., product, operator, and equipment, establishing the infrastructure for real-time changes and realizing dynamic characteristics. The authors used a hybrid deep neural network model based on the combination of MobileNetv2, YOLOv4, and OpenPose to further develop the learning algorithm to realize efficient multi-type small object detection.

1.4.3 Digital Twin Classifications

The digital twin is the most widely used virtual prototyping technology that acts as a bridge for integration between the physical and virtual components in terms of *object* and *process twinning*. Each type serves a distinct purpose, from improving product design and testing to enhancing system-wide operations and decision-making processes. This stratification allows for tailored solutions addressing specific challenges within various industries, from manufacturing to smart cities, enhancing efficiency, reliability, and innovation. More importantly, as the digitalization of industrial IoT becomes the foundation of smart manufacturing, DT is regarded as the biggest emerging technology trend and the most promising prototyping approach for realizing real-time object and process monitoring and maintenance in the industry. The digital twinning of objects aims at offering virtual counterparts of objects with real-time automated monitoring that takes part in automatic cyber-physical systems, while the digital twinning of a process virtualizes the whole CPS sub-system. Figure 1.5 shows the concept of object and process twinning from CPS to DT.

So far, object twinning has been driven by the concept of physics-based modeling methods that entail observing and understanding a physical phenomenon of the object, converting the comprehension into mathematical equations, and finally exploring them. Rasheed et al. [63], discussed several multi-physical simulation and physical realism

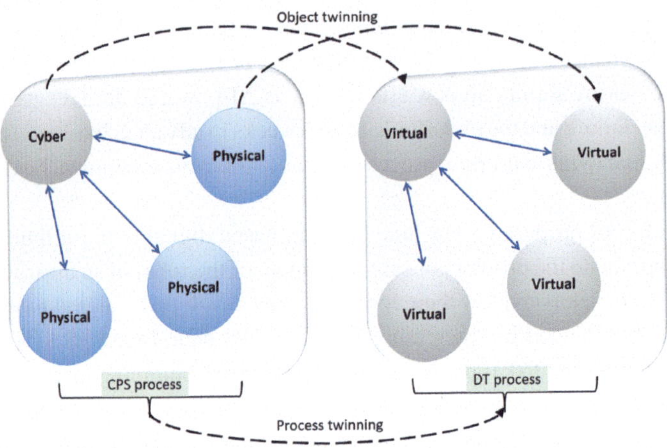

Fig. 1.5 Object and process twinning from CPS to DT

approaches e.g., Finite Difference Method (FDM), Finite Element Method (FEM), Finite Volume Method (FVM), and Discrete Element Method (DEM). Wu et al. [51] discussed in detail on physical object twinning, enabling technology for modeling, physical data processing, and network coordination among the physical and virtual twins in terms of network topology. More depictions of object twinning aimed at structural prediction and crack tip insights can be found mostly in the aviation domain (e.g., [70].

The process twining essentially includes the objects or components any process or system contains, though several DT structures do not necessarily virtualize the components i.e., they use an abstract form of each component and focus more on the sub-system computation. Alam et al. [10], introduced a digital twin architecture reference model for the cloud-based CPS, referred to as *C2PS*, where they analytically describe the key properties of their DT framework. *C2PS* allows exploring how the IoT sub-system can generate heterogeneous systems. They also validated the efficacy of their model by demonstrating a prototype driving assistance model for vehicular applications. Bao et al. [71] developed a DT architecture for the manufacturing sector that divides the DT into three components: product, process, and operation, each of which has a distinct architecture.

Several other researchers have classified DTs based on their functionality, i.e., as tools for simulation, facilitating virtual-real connectivity, building models, and processing data. Notably, Kritzinger et al. [72] made a distinction between digital models, digital shadows, and digital twins, each representing different levels of interaction between the physical and digital states. Digital models operate independently of their physical counterparts, digital shadows reflect changes in the physical object without affecting it, and digital twins offer a bidirectional interaction, where changes in one domain directly impact the other. This classification highlights the versatility of DTs in modeling and simulation, thereby enhancing our expertise in managing complex systems across diverse domains.

1.5 Enabling Technologies

Several enabling technologies are critical in building an intelligent virtual representation of a physical entity and supporting a continuous two-way feedback loop between the entities. In this section, we'll go through some of the most notable enabling technologies that researchers have used to develop VPS, DTs, or any virtualization frameworks to meet their needs. An overview of digital twin-related technologies outlined in various studies i.e., [73, 74] is shown in Fig. 1.6.

1.5.1 Modeling and Simulation

Digital twin frameworks for intelligent systems can be perceived as just another name for "simulation of your system design". The term "modeling and simulation" can refer to a wide range of frameworks and software components that can be used to direct developers in creating a critical component of a virtual representation of a physical object [75]. Therefore, modeling and simulation are key elements in building a virtual prototype. So, the question arises, "why use fancy names like digital twins and virtual platforms?" Although virtual prototypes and simulations both employ digital models to imitate items and processes, there are some significant differences between the two. If we look at the digital twin as an illustrative example—the most noteworthy aspect is that a digital twin develops a virtual

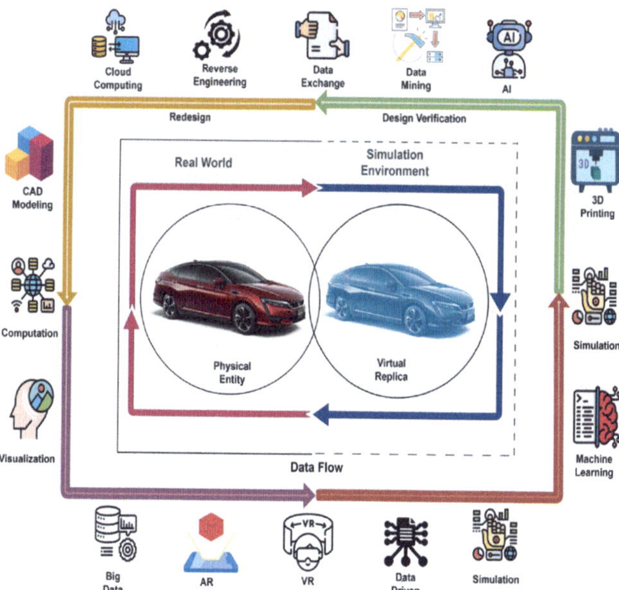

Fig. 1.6 Digital twin related technologies

environment capable of studying various simulations, backed by real-time data and a two-way communication channel between the sensor-equipped twin and the twin that collects the data. As a result, predictive analytical models become more accurate, providing better management and monitoring of products, regulations, and procedures [76].

The key differences between a digital twin and a simulation model can be seen in terms of the following categories.

- **Static versus Active**: A simulation model is static since it does not evolve unless the designer adds new components. While a digital twin will initially start off quite similar to a simulation model, the addition of real-time data allows the twin to alter and grow to provide a more active simulation [74]. A digital twin can develop over the course of a product lifecycle as more data is gathered and analyzed, providing unique comprehension that is not possible with a static simulation [77].
- **Present versus Predictive**: Simulations, at best, can assist in understanding what may occur in the real world. Digital twins help understand not only what might happen, but also what is happening [78].
- **Application**: A simulation is useful for product design because it enables designers to test various scenarios against predetermined criteria. The uses of a digital twin, however, are not restricted to specific business workflow areas because its scope is much broader and encompasses all stages of a product lifecycle.

With these distinctions in mind, a DT is developed from various modeling methodologies that provide the prototyping platform with the initial cyber backbone. From Fig. 1.1, we understand that VPs are traditionally developed using functional TLM models, system C, and OVP emulators i.e., building an abstraction level through these aforementioned modeling methodologies. The "Programmers' View" (PV) level of abstraction consists of the register-accurate transaction-level models (TLMs) of the peripherals and the instruction-set simulator (ISS) of the CPU [79]. The fundamental element of the TLM modeling approach is the distinction between the communication layer, functionality, and architectural elements as represented by time and power [80, 81].

In the case of digital twins, the modeling techniques are not specifically defined, as it is considered an interdisciplinary and versatile technology that varies by definition as well. Nonetheless, several researchers developed their own methods of modeling DT frameworks and attempted to establish it as an abstractly organized architecture. A group of authors [82, 83] established the modeling basis of the DT based on two components: a virtual twin of the physical entity and an API. They demonstrated the usage of API as a middleware that allows the DT to connect with external systems. Another research study explored the modeling techniques such as Design Elements, 8D-Model, and V-IoT to develop DT for smart factories [84]. In general, DTs are modeled from the characteristics of the target physical entity, and simulations are run based on the real-time behavioral data provided.

1.5.2 Digital Twin Feedback Loop

The effectiveness of a DT relies on a continuous sensor-driven feedback loop that enables real-time synchronization between the physical system and its virtual counterpart. This feedback loop begins with sensors deployed across the physical system to capture operational, environmental, and behavioral states. Depending on the application, these sensors may include electrical, mechanical, thermal, positional, or network-level monitors. Sensor measurements are streamed periodically or event-driven to the DT through industrial and IoT communication interfaces. Incoming sensor data undergoes preprocessing to address noise, latency, and heterogeneity across sensing modalities. This includes filtering, timestamp alignment, unit normalization, and feature extraction. The processed sensor data is then mapped to corresponding state variables and parameters within the DT, allowing the virtual model to track the evolving physical system with high fidelity. Once synchronized, the DT evaluates sensor-updated states using embedded analytical models. These models use physics-based relationships, data-driven predictors, or hybrid approaches to detect anomalies, assess system health, and forecast future behavior. The outcomes of these analyses form the basis of feedback decisions, such as control adjustments, fault mitigation actions, or optimization strategies.

Feedback is closed by translating DT decisions into control commands or recommendations that are applied to the physical system through actuators or supervisory control interfaces. The physical system's response is subsequently captured by sensors, initiating the next feedback cycle. This iterative process enables adaptive behavior, where the DT continuously refines its internal representation based on sensor evidence and system response. Through this closed-loop operation, sensor feedback not only reflects the instantaneous state of the physical system but also supports long-term model calibration and learning. As system dynamics evolve due to aging, workload variation, or environmental changes, sensor-informed updates ensure that the DT remains accurate and actionable over time.

1.5.3 Communications

One of the essential components of a virtual prototype is the communication framework that allows the physical and virtual entities to coexist. Additionally, the communication among the computation processes within a virtual prototype is crucial for the overall functionality of the cyber-physical system. To exchange data and information, the cooperating processes in a prototyping framework must communicate with one another. The mechanism for communicating between these processes is known as inter-process communication (IPC). The two modes of IPC are—shared memory and message passing (Fig. 1.7).

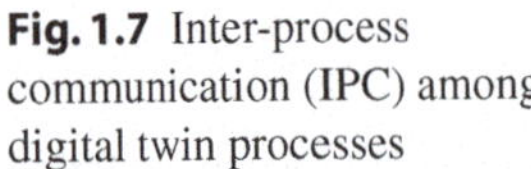

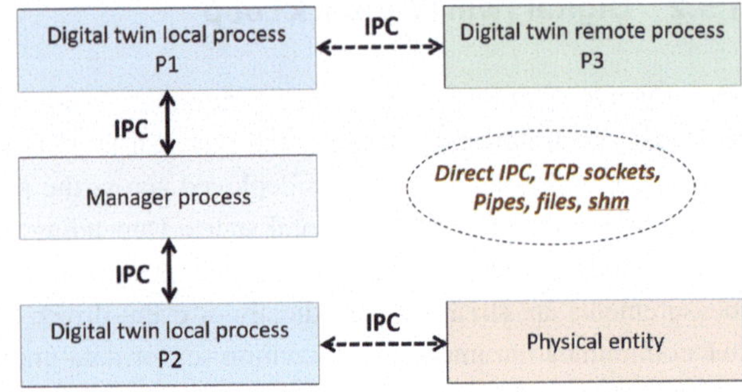

Fig. 1.7 Inter-process communication (IPC) among digital twin processes

DTs employ various kinds of IPC to establish communication in the virtual environment within those two IPC modes. Threads can now transcend process boundaries thanks to the OS feature known as "direct IPC," [85], which repurposes and expands the CODOMs [86] design. The OS kernel is removed from the critical inter-process communication path, and processes are mapped into a shared address space. Another useful communication feature is socket programming based on the client-server model [87]. These include various types of sockets, e.g., Transmission Control Protocol (TCP) [88], User Datagram Protocol (UDP), and raw sockets.

1.5.4 IoT Concepts

IoT envisions a world in which everything is intelligently connected, invoking technologies such as wireless sensor networks, digital twins, machine learning, edge computing, cloud computing, and many more within the context of distributed systems. The primary IoT concepts tell us about sensors, actuators, and controllers/systems embedded within the enabling technology. Modern virtual prototypes are often built based on this IoT concept, with the computational processes functioning as sensors, actuators, or controllers. IoT sensors (i.e., smart sensors) play a crucial role in sending real-time data to the virtual environment, allowing structural simulations. The depth of heterogeneous data that IoT sensors provide can be utilized to virtualize and visualize various industrial applications, enabling the prevention of risks and the remote management of workplace safety issues [89]. Figure 1.8 shows a generic conceptual framework for prototyping derived from various models proposed by academia and industry. Following an architectural study of some of the existing concepts, a set of layering principles and functionality is presented.

Edge and cloud computing are often used to move IoT sensory data across edge systems and public clouds. The virtualization of composite heterogeneous IoT systems always requires heavy processing, calling for the need for distributed computing. Jiang et al. [90] built a DT based on a traditional IoT framework that employs both edge and cloud computing, splitting the framework into two parts. Cloud platforms are more common for building

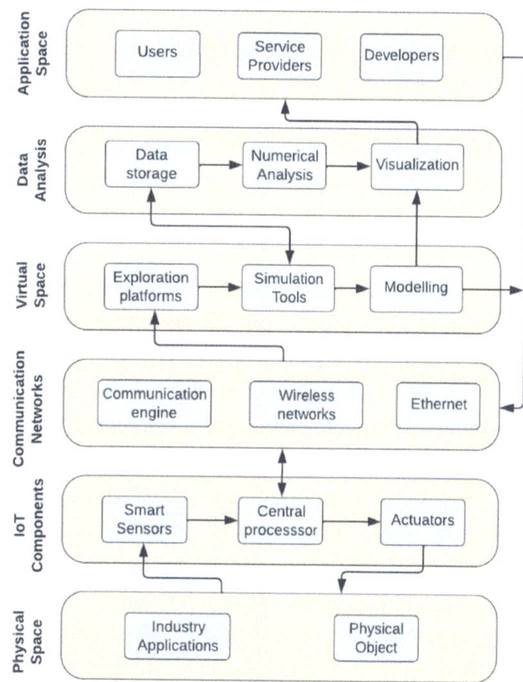

Fig. 1.8 Prototyping Conceptual framework within the Context of Internet-of-Things

DTs that take care of heavy computations while the prototyping framework synchronizes IoT-related functions between the physical and virtual entities [91, 92].

1.5.5 Virtual and Augmented Reality

Other than just computational and simulation-based models, prototypes are also constructed in a more visualized and immersive manner with the help of virtual reality (VR) and augmented reality (AR) technologies. VR is often used to navigate, interact with, and explore the virtual environment realistically, just like they would with the actual equipment. Users can learn about a certain physical entity immersively with a VR-based virtual environment without even interrupting the actual entity itself [93, 94]. AR helps to bring a DT component into the real environment by establishing on-demand live synchronization and integration. It can speed up access to virtual environment interfaces by overlaying virtual data and images on the camera feed while the camera is directed at the physical twin itself [95]. Microsoft HoloLens is widely used to visualize the AR components as part of the DT data in a real industrial environment [96].

Although both technologies have individually been crucial catalysts for prototyping activities, neither is sufficient on its own. Functionalities like two-way communications, real-time synchronization of virtual prototypes, and human-machine interfaces enable the virtual framework to achieve critical challenges. Having said that, some studies [97, 98] have combined both VR and AR technologies to achieve this goal, resulting in more efficient real-time simulation.

1.5.6 AI-Enabled and Generative Intelligence Technologies

Recent advances in AI, particularly in foundation models and large language models (LLMs), are increasingly influencing the development and operation of DTs. Unlike traditional machine learning techniques that are typically designed for narrow, task-specific objectives, modern generative AI models exhibit strong capabilities in representation learning, reasoning over heterogeneous data, and interacting with complex systems using natural language interfaces. These capabilities make them promising enablers for next-generation DT frameworks. One important role of AI-driven models in DTs is in enhancing data interpretation and decision support. DTs often ingest heterogeneous data streams originating from sensors, logs, simulation outputs, and historical records. LLMs and related generative models can assist in contextualizing this data by learning latent relationships across modalities and temporal scales. For example, AI models can help interpret sensor anomalies, correlate them with operational contexts, and generate human-readable explanations that support diagnostics and maintenance decisions .

Generative AI also enables more adaptive and self-evolving DT architectures. As physical systems evolve due to aging, configuration changes, or environmental variation, DT models require continuous refinement. AI-based learning mechanisms can support automated model calibration by adjusting parameters, updating behavioral models, or selecting appropriate simulation fidelities based on observed discrepancies between physical and virtual behavior. This is particularly useful in large-scale IoT and cyber-physical systems, where manual model maintenance becomes impractical [99, 100]. Another emerging capability is the use of LLMs as cognitive interfaces to DTs. Instead of interacting with DTs solely through dashboards or predefined APIs, users can query, control, and analyze DTs through natural language. Such interfaces can translate high-level user intent into low-level simulation commands, configuration updates, or analytical workflows. This paradigm significantly lowers the barrier to DT adoption for non-expert users and supports collaborative decision-making across interdisciplinary teams.

AI-driven generative models further facilitate scenario generation and what-if analysis. By learning from historical data and simulation outcomes, these models can synthesize plausible future scenarios, stress-test system behavior under rare conditions, and support proactive risk assessment. In combination with reinforcement learning and optimization techniques, DTs can leverage AI to recommend control actions or operational strategies that improve efficiency, resilience, or safety. Despite these advantages, integrating generative AI into DTs introduces new challenges, including model trustworthiness, explainability, computational overhead, and data governance. Ensuring that AI-generated insights are interpretable and aligned with physical constraints remains critical, particularly in safety- and mission-critical domains. Consequently, current research increasingly views AI not as a replacement for physics-based or system models, but as a complementary layer that augments DT intelligence, adaptability, and usability.

1.6 Conclusion

The Internet of Things is ushering in a confluence of many disciplines, including artificial intelligence, computing system design, and smart industrial infrastructure. In the past decade, virtual prototyping solutions for various IoT applications have advanced in the form of virtual platforms, digital twins, a blend of both, and concepts not analogous to either. We presented a comprehensive high-level overview of virtual prototyping solutions in IoT applications. We discussed how this technology has proven beneficial in three key aspects i.e., system exploration, optimization, and security. This chapter has addressed and disambiguated three critical issues: (1) the definitions and types of various virtual prototyping techniques in industrial IoT; (2) the characteristics of various prototyping solutions in terms of modeling, simulation, and application; and (3) the enabling technologies that are critical in building digital twins in modern IoT industry. We believe that the overview will pave the way for a comprehensive understanding of the state of the art in this area and facilitate future research. In particular, novel architectural concepts are being developed, especially in the area of digital twins. Therefore, robust and hybrid forms of virtualization are combining various modeling approaches that can become more appealing to the development of more intelligent, secure, and autonomous systems.

References

1. Kosmatos Evangelos, A., D. Tselikas Nikolaos, and C. Boucouvalas Anthony. 2011. Integrating RFIDs and smart objects into a UnifiedInternet of Things architecture. *Advances in Internet of Things* 2011.
2. Hu, Junyan, Barry Lennox, and Farshad Arvin. 2022. Robust formation control for networked robotic systems using Negative Imaginary dynamics. *Automatica* 140:110235.
3. Shafi, Qaisar. 2012. Cyber physical systems security: A brief survey. In *2012, 12th International Conference on Computational Science and Its Applications*, 146–150. IEEE.
4. Poovendran, Radha. 2010. Cyber–physical systems: Close encounters between two parallel worlds [point of view]. *Proceedings of the IEEE* 98(8): 1363–1366.
5. Shafique, Kinza, et al. 2020. Internet of things (IoT) for next-generation smart systems: A review of current challenges, future trends and prospects for emerging 5G-IoT scenarios. *IEEE Access* 8: 23022–23040.
6. Boyes, Hugh, et al. 2018. The industrial internet of things (IIoT): An analysis framework. *Computers in Industry* 101: 1–12.

7. Kabir, Md. Rafiul. 2024. *Digital Twin Exploration Platforms for IoT Applications of High Global Impact*. University of Florida.
8. Kabir, Md Rafiul, and Sandip Ray. 2023. Virtual prototyping for modern internet-of-things applications: A survey. *IEEE Access* 11: 31384–31398.
9. Mueller, Wolfgang et al. 2012. Virtual prototyping of cyber-physical systems. In *17th Asia and South Pacific Design Automation Conference*, 219–226. IEEE.
10. Kazi Masudul Alam and Abdulmotaleb El Saddik. 2017. C2PS: A digital twin architecture reference model for the cloud-based cyber-physical systems. *IEEE Access* 5: 2050–2062.
11. Lee, Jay, Behrad Bagheri, and Hung-An Kao. 2015. A cyber-physical systems architecture for industry 4.0-based manufacturing systems. *Manufacturing Letters* 3: 18–23.
12. Jia, Dongyao et al. 2015. A survey on platoon-based vehicular cyber-physical systems. *IEEE Communications Surveys & Tutorials* 18(1): 263–284.
13. Yan, Xu., et al. 2017. Industrial big data for fault diagnosis: Taxonomy, review, and applications. *IEEE Access* 5: 17368–17380.
14. Catarci, Tiziana et al. 2019. A conceptual architecture and model for smart manufacturing relying on service-based digital twins. In *2019 IEEE International Conference on Web Services (ICWS)*, 229–236. https://doi.org/10.1109/ICWS.2019.00047
15. Zhou, Yimin et al. 2019. Video coding optimization for virtual reality 360-degree source. *IEEE Journal of Selected Topics in Signal Processing* 14(1): 118–129.
16. Schuster, Thomas. 2014. SoCRocket-A virtual platform for the European Space Agency's SoC development. In *2014, 9th International Symposium on Reconfigurable and Communication-Centric Systems-on-Chip (ReCoSoC)*, vol. 2014, 1–7. IEEE.
17. Yeh, Tse-Chen, Zin-Yuan Lin, and Ming-Chao Chiang. 2010. Optimizing the simulation speed of qemu and systemc-based virtual platform. In *2010 2nd International Conference on Information Engineering and Computer Science*, 1–4. IEEE.
18. Hawilo, Hassan, Manar Jammal, and Abdallah Shami. 2017. Orchestrating network function virtualization platform: Migration or re-instantiation? In *2017 IEEE 6th International Conference on Cloud Networking (CloudNet)*, 1–6. IEEE.
19. Wang, Luhan, et al. 2016. Joint optimization of service function chaining and resource allocation in network function virtualization. *IEEE Access* 4: 8084–8094.
20. Jeon, S.M., and Sebastian Schuesslbauer. 2020. Digital twin application for production optimization. In *2020 IEEE International Conference on Industrial Engineering and Engineering Management (IEEM)*, 542–545. IEEE.
21. Vachálek, Ján. 2017. "The digital twin of an industrial production line within the industry 4.0 concept. In *2017 21st International Conference on Process Control (PC)*, 258–262. IEEE.
22. Stojanovic, Nenad, and Dejan Milenovic. 2018. Data-driven Digital Twin approach for process optimization: An industry use case. In *2018 IEEE International Conference on Big Data (Big Data)*, 4202–4211. IEEE.
23. Karnouskos, Stamatis. 2011. Stuxnet worm impact on industrial cyber-physical system security. In *IECON 2011—37th Annual Conference of the IEEE Industrial Electronics Society*, 4490–449. https://doi.org/10.1109/IECON.2011.6120048.
24. He, Hongmei. 2016. The security challenges in the IoT enabled cyber-physical systems and opportunities for evolutionary computing & other computational intelligence. In *2016 IEEE Congress on Evolutionary Computation (CEC)*, vol. 2016, 1015–1021. IEEE.
25. Kulik, Tomas et al. 2022. Towards secure digital twins. In *Leveraging Applications of Formal Methods, Verification and Validation. Practice: 11th International Symposium, ISoLA 2022, Rhodes, Greece, October 22–30, 2022, Proceedings, Part IV*, 159–176. Springer.
26. Almeaibed, Sadeq et al. 2021. Digital twin analysis to promote safety and security in autonomous vehicles. *IEEE Communications Standards Magazine* 5(1): 40–46.

27. Alcaraz, Cristina, and Javier Lopez. 2022. Digital twin: A comprehensive survey of security threats. *IEEE Communications Surveys & Tutorials.*
28. Lou, Xinxin et al. 2019. An idea of using Digital Twin to perform the functional safety and cybersecurity analysis. In *INFORMATIK 2019: 50 Jahre Gesellschaft f ü r Informatik–Informatik f ü r Gesellschaft (Workshop-Beitr ä ge).* Gesellschaft für Informatik eV.
29. Ahn, Sunha, and Sharad Malik. 2014. Automated firmware testing using firmware-hardware interaction patterns. *CODES+ISSS* 25: 1–25:10.
30. Kannavara, R. et al. 2015. Challenges and opportunities in concolic testing. *Proceedings of the National Aerospace Electronics Conference—Ohio Innovation Summit.*
31. Leupers, Rainer. 2012. Virtual platforms: Breaking new grounds. 2012 Design. *Automation & Test in Europe Conference & Exhibition (DATE)*, 685–669. https://doi.org/10.1109/DATE.2012.6176558.
32. Kun, Lu, Daniel Müller-Gritschneder, Ulf Schlichtmann. 2012. Accurately timed transaction level models for virtual prototyping at high abstraction level. *2012 Design, Automation & Test in Europe Conference & Exhibition (DATE)*, 135–140. IEEE.
33. Biondi, Alessandro, et al. 2021. SPHERE: A Multi-SoC architecture for next-generation cyber-physical systems based on heterogeneous platforms. *IEEE Access* 9: 75446–75459.
34. Hong, Sungpack et al. 2006. Creation and utilization of a virtual platform for embedded software optimization: An industrial case study. In *Proceedings of the 4th International Conference on Hardware/Software Codesign and System Synthesis*, 235–240.
35. Lin, Yi.-Li., W. YSu. Alvin. 2011. Functional verifications for SoC software, hardware co-design: From virtual platform to physical platform. In *2011 IEEE International SoC Conference*, 201–206. IEEE.
36. Chiang, Ming-Chao, Tse-Chen Yeh, and Guo-Fu Tseng. 2011. A QEMU and SystemC-based cycle-accurate ISS for performance estimation on SoC development. In *IEEE Transactions on Computer-Aided Design of Integrated Circuits and Systems*, vol. 30.4, 593–606.
37. Cucchetto, Filippo, Alessandro Lonardi, and Graziano Pravadelli. 2014. A common architecture for co-simulation of SystemC models in QEMU and OVP virtual platforms. In *2014 22nd International Conference on Very Large Scale Integration (VLSI-SoC)*, 1–6. IEEE.
38. Peng, Cheng-Shiuan. 2010. Dual-core virtual platform with QEMU and System. In *2010 International Symposium on Next Generation Electronics*, 69–72. IEEE.
39. Delbergue, Guillaume et al. 2016. QBox: An industrial solution for virtual platform simulation using QEMU and SystemC TLM-2.0. In *8th European Congress on Embedded Real Time Software and Systems (ERTS 2016).*
40. Vora, M., Priyabarata Mishra, and Anirudha Shrikant Kurhade. 2022. Integration of Pycachesim with QEMU. In *2021 4th International Conference on Recent Trends in Computer Science and Technology (ICRTCST)*, 27–30. IEEE.
41. Wicaksana, Arya, and Chong Ming Tang. 2017. Virtual prototyping platform for multiprocessor system-on-chip hardware/software co-design and co-verification. In *International Conference on Computer and Information Science*, 93–108. Springer.
42. Ahn, SungWon et al. 2015. Implementing virtual platform for global-scale cyber physical system networks. *International Journal of Communication Systems* 28(13): 1899–1920.
43. Soares, Joao. 2014. Cloud4nfv: A platform for virtual network functions. In *2014 IEEE 3Rd International Conference on Cloud Networking (cloudnet)*, 288–293. IEEE.
44. Monteleone, Giuseppe, and Pietro Paglierani. 2013. Session border controller virtualization towards "Service-Defined" networks based on NFV and SDN. In *2013 IEEE SDN for Future Networks and Services (SDN4FNS)*, 1. https://doi.org/10.1109/SDN4FNS.2013.6702554.
45. Batalle, Josep, 2013. On the implementation of NFV over an OpenFlow infrastructure: Routing function virtualization. In *2013 IEEE SDN for Future Networks and Services (SDN4FNS)*, 1–6. IEEE.

46. Wagg, D. J. et al. 2020. Digital Twins: State-of-the-Art and future directions for modeling and simulation in engineering dynamics applications. *ASCE-ASME Journal of Risk and Uncertainty in Engineering Systems, Part B: Mechanical Engineering* 6(3).
47. Grieves, Michael W. 2019. Virtually intelligent product systems: Digital and physical twins.
48. Qi, Qinglin, et al. 2021. Enabling technologies and tools for digital twin. *Journal of Manufacturing Systems* 58: 3–21.
49. Barricelli, Barbara Rita, Elena Casiraghi, and Daniela Fogli. 2019. A survey on digital twin: Definitions, characteristics, applications, and design implications. *IEEE Access* 7: 167653–167671.
50. Boschert, Stefan, and Roland Rosen. 2016. Digital twin—the simulation aspect. In *Mechatronic Futures*, 59–74. Springer.
51. Wu, Yiwen, Ke Zhang, and Yan Zhang. 2021. Digital twin networks: A survey. *IEEE Internet of Things Journal*.
52. Negri, Elisa, Luca Fumagalli, and Marco Macchi. 2017. A review of the roles of digital twin in CPS-based production systems. *Procedia Manufacturing* 11: 939–948.
53. *Digital Twin: Siemens*. 2021. https://www.plm.automation.siemens.com/global/en/our-story/glossary/digital-twin/24465.
54. Madni, Azad M., Carla C Madni, and Scott D. Lucero. 2019. Leveraging digital twin technology in model-based systems engineering. *Systems* 7(1): 7.
55. Gabor, Thomas, 2016. A simulation-based architecture for smart cyber-physical systems. In *2016. IEEE International Conference on Autonomic Computing (ICAC)*, 374–379. IEEE.
56. Qi, Qinglin, and Fei Tao. 2018. Digital twin and big data towards smart manufacturing and industry 4.0: 360 degree comparison. *IEEE Access* 6: 3585–3593.
57. Grieves, Michael, and John Vickers. 2017. Digital twin: Mitigating unpredictable, undesirable emergent behavior in complex systems. In *Transdisciplinary Perspectives on Complex Systems*, 85–113. Springer.
58. Boschert, Stefan, Christoph Heinrich, and Roland Rosen. 2018. Next generation digital twin. In *Proceedings of the tmce*, vol. 2018, 7–11. Spain: Las Palmas de Gran Canaria.
59. Tao, Fei, and Meng Zhang. 2017. Digital twin shop-floor: A new shop-floor paradigm towards smart manufacturing. *IEEE Access* 5: 20418–20427.
60. Tao, Fei et al. 2018. Digital twin driven prognostics and health management for complex equipment. *Cirp Annals* 67(1): 169–172.
61. Goodfellow, Ian et al. 2020. Generative adversarial networks. *Communications of the ACM* 63(11): 139–144.
62. Fischer, Asja, and Christian Igel. 2012. An introduction to restricted Boltzmann machines. In *Progress in Pattern Recognition, Image Analysis, Computer Vision, and Applications: 17th Iberoamerican Congress, CIARP 2012, Buenos Aires, Argentina, 3–6 Sept 2012. Proceedings 17*, 14–36. Springer.
63. Rasheed, Adil, Omer San, and Trond Kvamsdal. 2020. Digital twin: Values, challenges and enablers from a modeling perspective. *IEEE Access* 8: 21980–22012.
64. Nguyen, Huan X. et al. 2021. Digital Twin for 5G and Beyond. *IEEE Communications Magazine* 59(2): 10–1. https://doi.org/10.1109/MCOM.001.2000343.
65. Wärmefjord, Kristina et al. 2017. Inspection data to support a digital twin for geometry assurance. In: *ASME International Mechanical Engineering Congress and Exposition*, Vol. 58356. American Society of Mechanical Engineers. V002T02A101.
66. Dröder, Klaus, et al. 2018. A machine learning-enhanced digital twin approach for human-robot-collaboration. *Procedia Cirp* 76: 187–192.
67. Jain, Anil K., Jianchang Mao, and K Moidin Mohiuddin. 1996. Artificial neural networks: A tutorial. *Computer* 29(3): 31–44.

68. Hürkamp, André et al. 2020. Combining simulation and machine learning as digital twin for the manufacturing of overmolded thermoplastic composites. *Journal of Manufacturing and Materials Processing* 4(3): 92.
69. Zhou, Xiaokang et al. 2021. Intelligent small object detection based on digital twinning for smart manufacturing in industrial CPS. *IEEE Transactions on Industrial Informatics*.
70. Majumdar, Prasun K., Mohammad FaisalHaider, and Kenneth Reifsnider. 2013. Multi-physics response of structural composites and framework for modeling using material geometry. In *54th AIAA/ASME/ASCE/AHS/ASC Structures, Structural Dynamics, and Materials Conference*, 1577.
71. Bao, Jinsong et al. 2019. The modelling and operations for the digital twin in the context of manufacturing. *Enterprise Information Systems* 13(4): 534–556.
72. Kritzinger, Werner et al. 2018. Digital Twin in manufacturing: A categorical literature review and classification. *Ifac-PapersOnline*, 51(11): 1016–1022.
73. Lo, C.K., C.H. Chen, and Ray Y. Zhong. 2021. A review of digital twin in product design and development. *Advanced Engineering Informatics* 48: 101297.
74. Tao, Fei et al. 2019. Digital twin-driven product design framework. *International Journal of Production Research* 57(12): 3935–3953.
75. Mihai, Stefan et al. 2022. Digital twins: A survey on enabling technologies, challenges, trends and future prospects. *IEEE Communications Surveys & Tutorials*.
76. *Simulation vs Digital Twin (what is the difference between them?)*. https://www.twi-global.com/technical-knowledge/faqs/simulation-vs-digital-twin.
77. Wright, Louise, and Stuart Davidson. 2020. How to tell the difference between a model and a digital twin. *Advanced Modeling and Simulation in Engineering Sciences* 7(1): 1–13.
78. Raghunathan, Vijay. 2019. *Digital Twins vs Simulation: Three key differences*, May 2019. https://www.entrepreneur.com/en-au/technology/digital-twins-vs-simulation-three-key-differences/333645.
79. Sisodia, Umesh. 2011. *Creating an SOC virtual platform for embedded software development* ..., June 2011. https://www.electronicdesign.com/news/products/article/21759920/creating-an-soc-virtual-platform-for-embedded-software-development.
80. Wang, Chen-Chieh. 2009. System-level development and verification framework for high-performance system accelerator. In *2009 International Symposium on VLSI Design, Automation and Test*, 359–362. IEEE.
81. Magdy, A., El-Moursy. 2014. Efficient embedded SoC hardware, software codesign using virtual platform. In *2014 9th International Design and Test Symposium (IDT)*, 36–38. IEEE.
82. Schroeder, Greyce N. et al. 2016. Digital twin data modeling with automationml and a communication methodology for data exchange. *IFAC-PapersOnLine* 49(30): 12–17.
83. Schroeder, Greyce N. et al. 2020. A methodology for digital twin modeling and deployment for industry 4.0. *Proceedings of the IEEE*, 109(4): 556–567.
84. Riedelsheimer, Theresa, Sonika Gogineni, and Rainer Stark. 2021. Methodology to develop Digital Twins for energy efficient customizable IoT-Products. *Procedia CIRP* 98: 258–263.
85. Vilanova, Lluís et al. 2017. Direct Inter-Process Communication (dIPC) Repurposing the CODOMs Architecture to Accelerate IPC. In *Proceedings of the Twelfth European Conference on Computer Systems*, 16–31.
86. Vilanova, Lluís et al. 2014. CODOMs: Protecting software with code-centric memory domains. *ACM SIGARCH Computer Architecture News* 42(3): 469–480.
87. Xue, Ming, and Changjun Zhu. 2009. The socket programming and software design for communication based on client/server. In *2009 Pacific-Asia Conference on Circuits, Communications and Systems*, 775–77. https://doi.org/10.1109/PACCS.2009.89.
88. Postel, Jon. 1981. *Transmission control protocol*. Technical report.

89. Zhao, Zhiheng et al. 2021. IoT and digital twin enabled smart tracking for safety management. *Computers & Operations Research* 128: 105183.
90. Jiang, Zongmin, Yangming Guo, and Zhuqing Wang. 2021. Digital twin to improve the virtual-real integration of industrial IoT. *Journal of Industrial Information Integration* 22: 100196.
91. Wang, Ziran, 2020. A digital twin paradigm: Vehicle-to-cloud based advanced driver assistance systems. In *2020 IEEE 91st Vehicular Technology Conference (VTC2020-Spring)*, 1–6. IEEE.
92. Wenjun, Xu., et al. 2021. Digital twin-based industrial cloud robotics: Framework, control approach and implementation. *Journal of Manufacturing Systems* 58: 196–209.
93. Kaarlela, Tero, Sakari Pieskä, and Tomi Pitkäaho. 2020. Digital twin and virtual reality for safety training. In *2020 11th IEEE International Conference on Cognitive Infocommunications (CogInfoCom)*, 000115–000120. IEEE.
94. Karadeniz, Ahmet Mert. 2019. Digital twin of eGastronomic things: A case study for ice cream machines. In *2019 IEEE International Symposium on Circuits and Systems (ISCAS)*, 1–4. IEEE.
95. Schroeder, Greyce, 2016. Visualising the digital twin using web services and augmented reality. In *2016 IEEE 14th international conference on industrial informatics (INDIN)*, 522–527. IEEE.
96. Zhou, Mike, Jianfeng Yan, and Donghao Feng. 2019. Digital twin framework and its application to power grid online analysis. *CSEE Journal of Power and Energy Systems* 5(3): 391–398.
97. Qiu, Chan et al. 2019. Digital assembly technology based on augmented reality and digital twins: A review. *Virtual Reality & Intelligent Hardware* 1(6): 597–610.
98. Ma, Xin, et al. 2019. Digital twin enhanced human-machine interaction in product lifecycle. *Procedia Cirp* 83: 789–793.
99. Tao, Fei et al. 2019. Digital twin-driven product design framework. *International Journal of Production Research* 57(12): 3935–3953.
100. Rathore, M. Mazhar et al. 2021. The role of AI, machine learning, and big data in digital twinning: A systematic literature review, challenges, and opportunities. *IEEE Access* 9(3): 2030–32052.

Digital Twin Applications in IoT-Based Energy Systems

2

Digital twins driving the evolution of intelligent and resilient energy ecosystems.

2.1 Introduction

The global energy sector has been moving toward a more digitalized and interconnected future in recent times because of the advancement of technologies like the Internet of Things, cloud computing, and machine learning/artificial intelligence [1]. In the evolving landscape of energy systems, digital twin technology has emerged as a pivotal innovation, enabling optimization, efficiency, and resilience. The adoption of this technology in smart energy systems has been driven by its capacity to address complex challenges such as demand forecasting, asset management, and the integration of renewable energy sources, ultimately contributing to the sustainability and reliability of energy infrastructures [2]. Significantly, the implementation of digital twins has been identified as a catalyst for enhancing operational performance, extending asset lifespans, and ensuring compliance with stringent environmental standards. These virtual models serve not only to validate and monitor real-time functions but also to simulate and optimize processes before physical implementation, thereby reducing the need for costly prototypes and expediting development cycles. The integration of digital twins with advanced technologies such as artificial intelligence (AI), the Internet of Things (IoT), and cloud computing has further augmented their capabilities, enabling energy systems to adapt to rapidly changing conditions and demands.

The rapid growth in digital twins, with an expected market increase of 42.7% between 2021 and 2028, highlights their key role in transitioning energy systems to be more agile, resilient, and sustainable [3]. To achieve this transformation, IoT technology is crucial for making energy infrastructures smarter and more automated. IoT facilitates real-time

M. R. Kabir and S. Ray, *Digital Twins for Distributed IoT Applications*, Synthesis Lectures on Engineering, Science, and Technology, https://doi.org/10.1007/978-3-032-18938-7_2

monitoring and control of energy resources, enabling the dynamic optimization of energy production, distribution, and consumption. By embedding sensors and smart devices throughout the energy network, IoT allows for the seamless collection of data on energy usage patterns, system performance, and environmental conditions. This data-driven intelligence supports the development of smart grids that can automatically balance supply and demand, integrate renewable energy sources more effectively, and enhance the reliability and resilience of energy systems. Furthermore, IoT applications empower consumers with detailed insights into their energy consumption, fostering energy-saving behaviors and contributing to the overall reduction of carbon footprints.

In the framework of IoT-driven energy systems, this chapter provides a thorough analysis of digital twin technology, evaluating its impact, challenges, and prospects for advancement. Several studies have focused on generic energy systems or specific sections of the energy infrastructure. However, our study specifically focused on the applications of DTs solely within the energy infrastructure that is enhanced with IoT technology. We will explore the practical applications of digital twins in this specialized field by reviewing a variety of studies, with a particular emphasis on innovative approaches and advanced tools that are shaping the future of sustainability and energy optimization. Since this is primarily a survey, we do not delve deeply into the technical architectures, models, or optimization tools of specific DTs. Given the breadth of research in this area, covering all technical models in detail would be challenging. We showcase the creative approaches and fixes that IoT-driven systems provide to improve energy conservation and management through this particular lens. Figure 2.1 presents a taxonomy of digital twins in smart energy systems discussed in this chapter. While it does not directly address specific energy generation sources, it is important to note that digital twins can be employed across a variety of energy sources—

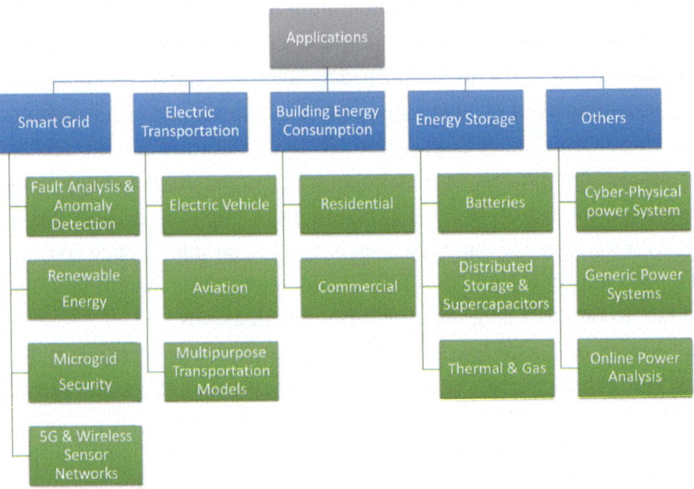

Fig. 2.1 A taxonomy of digital twins in smart energy systems

renewable and non-renewable—to enhance system efficiency, reliability, and integration into the energy mix. Their role in balancing and optimizing the energy mix is critical, particularly in facilitating the transition to a more sustainable energy landscape by improving the performance of both renewable energy sources and conventional power generation. While the diagrams throughout the chapter share a common theme, i.e., demonstrating how digital twins can be applied across various domains—the aim is to highlight the diversity of applications enabled by it. The recurring theme across the diagrams emphasizes that, despite the varied applications, the technology consistently enhances system performance across different fields.

2.2 Background

2.2.1 Digital Twin Concept in Energy Systems

Several studies have coined the term "Energy Digital Twin" (EDT) [4, 5]—defined as a digital representation that mimics the behavior and performance of physical energy systems. Therefore, in the context of energy systems, an EDT can represent a range from a single wind turbine to an entire power grid, incorporating renewable energy sources, power plants, transmission lines, and distribution networks.

2.2.2 IoT-Driven Energy

The term "IoT-driven energy system" refers to the integration of the Internet of Things technology into the management and operation of energy systems. IoT enables a network of interconnected devices to collect, exchange, and process data without the need for human-to-computer or human-to-human interaction. When applied to energy systems, IoT technology facilitates the monitoring, control, and optimization of energy production, distribution, and consumption in a more efficient, reliable, and sustainable manner. Here is a discussion of its meaning and implications:

- Monitoring and Data Collection: IoT devices, such as smart meters and sensors, collect data on energy usage, environmental conditions, and system performance in real-time [6]. This data is crucial for understanding patterns, identifying inefficiencies, and making informed decisions.
- Control and Automation: Through IoT, energy systems can automatically adjust based on the data received. For instance, temperature settings on smart thermostats can be changed in response to occupancy and the climate, and smart grids can balance supply and demand across the network.

- Optimization: IoT-driven energy systems can analyze data from various sources to optimize energy consumption and reduce waste. For instance, by predicting peak demand times, an IoT-enabled system can store energy during off-peak times and release it during peak times, thereby enhancing energy efficiency.
- Integration of Renewable Energy: IoT is essential in more efficiently incorporating renewable energy sources, like solar and wind, into the current power grid. It allows for real-time monitoring and control of energy production, enabling a more seamless shift towards cleaner energy solutions [7].
- Enhanced Reliability and Maintenance: With IoT, energy systems can predict and identify potential issues before they cause system failures. Predictive maintenance minimizes downtime and extends the lifespan of equipment by scheduling maintenance based on actual needs rather than fixed intervals.
- User Engagement and Demand Response: IoT-driven systems can engage consumers by providing them with detailed energy usage data, enabling them to make more informed decisions about their consumption. Additionally, demand response programs can be implemented more effectively, where consumers are motivated to reduce or shift their energy use during peak times [8].

In summary, IoT is revolutionizing existing energy systems, transforming them into smarter, more secure infrastructures. This transformation has inspired us to concentrate exclusively on the DTs of smart energy systems, analyzing both their current applications and future research directions, as well as their potential impacts.

2.2.3 Related Surveys and Reviews

Numerous reviews discuss the application of DTs across various energy systems, ranging from comprehensive overviews to those concentrating on specific aspects of energy, such as smart grids and energy storage. It is important to note that this section is not designed to comprehensively review all existing literature in the field but instead aims to provide an overview of prior research and perspectives on the topic, guiding the development of the research protocol. Consequently, some keywords that might be synonymous with the terms adopted were not considered during this exploratory phase. We summarize these papers in Table 2.1.

Table 2.1 Summary of related surveys and reviews on DTs in energy

Topic	Keywords	Reference
Comprehensive review	Digital twin, Energy systems, IoT, Energy optimization, Planning and operations, Digitalization, Energy supply, Energy demand, Energy storage, Decision-making Renewable energy, Literature review	Do Amaral et al. [5], Ghenai et al. [9]
Smart manufacturing and energy management	Digital Twin, Smart grid, Industry 4.0, Smart energy application, Modelling, Low carbon city, Electrified transportation, Energy storage system, Barriers, Sustainable energy, Energy engineering, Process systems engineering	Wang et al. [10], Lamagna et al. [11], Yu et al. [4]
Energy storage	Digital twin, Energy storage, Optimization, Electric vehicle, Cyber-physical systems, Batteries	Semeraro et al. [12], Vandana et al. [13]
Power electronics-based energy conversion systems	Not specified	Chen et al. [14]
Smart grid, smart city, and transportation	Transportation system, Digital twin, Microgrid, Security Physical twin, Machine learning, Power system	Jafari et al. [15]
Electric vehicle	Digital twin, IoT, Power converter Vehicle health monitoring, Battery management system, Sustainable transportation, Smart electric vehicles, Intelligent charging	Bhatti et al. [16]
Power system research and development	Digital twin, Communication channel, Machine learning, Internet of things, Cyber-physical systems	Yassin et al. [17]
Electric grid	Not specified	Sifat et al. [18]
IoT-driven energy systems	Digital twin, Energy, Internet of Things, Smart grid	Kabir et al. [19]
Building energy consumption	Operation, Maintenance, Digital twin, IoT system, Building Information, BIM, Energy efficiency, Energy modeling	Cespedes-Cubides et al. [20]

2.3 DT Application for Smart Energy Systems

We explore a range of digital applications within smart energy systems, focusing on the enhancement of smart grids, energy consumption, and storage, as well as the advancement of electric transportation and other related energy systems. This categorization facilitates a targeted analysis of each application's unique contributions to energy efficiency and sustainability, thereby streamlining the development of effective DT solutions. Indeed, some areas naturally intersect. For example, Vehicle-to-Grid (V2G) technology can be part of both smart grids and electric transportation, while energy storage can span across smart grids and building management. While we have categorized these areas separately for clarity, some overlap is inevitable due to the integrated nature of modern energy systems. Our structure aims to reflect the key aspects of each area while recognizing these interdependencies.

2.3.1 Smart Grids

The evolution of smart grids marks a significant leap in the utility industry's adoption of advanced technologies to enhance efficiency, reliability, security, and service quality. By leveraging innovations such as cloud computing, reinforcement learning, and big data analytics, smart grids are transforming energy networks through improved communication, automation, and connectivity [21, 22]. These grids are not only integrating both non-renewable and renewable energy sources to meet growing energy demands but also focusing on minimizing environmental impacts and promoting sustainability [23]. The drive towards smarter energy solutions underscores the critical role of communication and information technology in overcoming challenges brought on by the growing size and complexity of power management, environmental concerns, and the quest for energy independence and superior service quality [24]. This transformation represents a pivotal shift towards utilizing emerging technologies e.g., digital twins, to foster a more efficient, secure, and sustainable energy future.

Smart grids exemplify the best representation of IoT-based energy applications, and to better understand their scope, we categorize these applications based on the use of digital twins. This division helps in exploring the diverse functionalities and efficiencies these systems bring to energy management and distribution. Figure 2.2 provides an overview of a DT for a smart grid, comprising a grid with all IoT components, and detailing energy distribution and consumption.

2.3.1.1 Fault Analysis and Anomaly Detection

Digital twin technology is increasingly used for fault analysis and anomaly detection in smart grids, offering a practical solution to improve grid reliability and efficiency through virtual replicas that simulate and analyze grid operations in real-time. Tzanid et al. [25] addresses the increased complexity in managing and operating power networks due to the

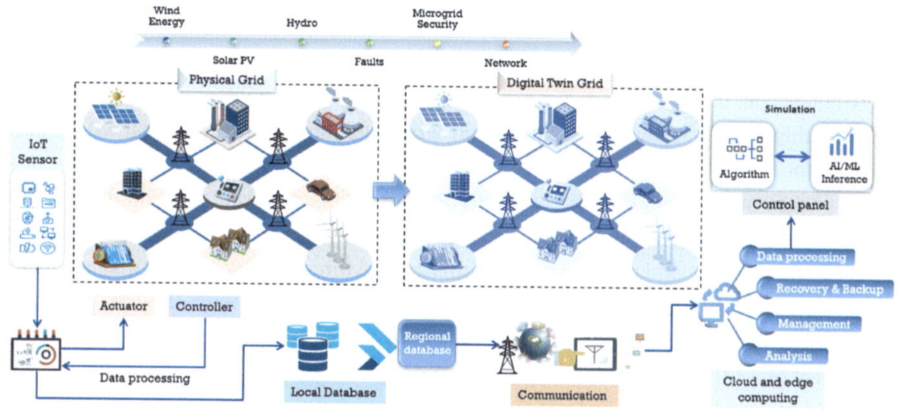

Fig. 2.2 An overview of digital twin for smart grid

widespread integration of high-speed power electronics and Distributed Energy Resources (DER), especially at the distribution grid edge. It introduces a CPS model for a distribution subsystem, enhanced by a low latency Industrial Internet of Things (IIoT) network and edge computing resources, including a DT represented by a hybrid model that combines a data-driven machine learning approach with a discrete-time model-based transient state estimator. Another work [26] proposes a deep learning convolutional neural network (CNN) approach within the DT environment for detecting and classifying physical faults in power systems. Highlighting the critical nature of the power grid and its evolution into a complex CPS, the method leverages high-fidelity measurement data from IEEE benchmark power systems to ensure reliability and protect infrastructure by accurately identifying fault locations. Olatunji et al. [27] highlighted the significance of DT technology for fault diagnosis and condition monitoring in wind turbines, presenting an overview of its application in identifying and managing mechanical component failures. The benefits, challenges, and necessary steps to fully harness the potential of DTs in the wind energy industry are discussed, emphasizing the need for enhanced accessibility, visibility, and availability.

In the photovoltaic domain, Jain et al. [28] developed a DT for real-time fault diagnosis in PV energy conversion units, incorporating a source-level power converter and demonstrating the detection and identification of various faults with high sensitivity. Their experimental validation showed the ability to identify faults in the power converter and sensors under 290 microseconds and faults in PV panels in less than 80 milliseconds, outperforming existing methods. Furthermore, several works [26, 29], [30] have made significant contributions to anomaly detection in the power grid using deep learning and supervised learning models.

2.3.1.2 Renewable Energy

Recent applications of renewable energy in smart grids have significantly transformed the energy landscape, making power systems more sustainable, efficient, and resilient. Integrating renewable energy sources (e.g., solar and wind) into smart grids has enabled dynamic management of both supply and demand, using advanced technologies for real-time monitoring, energy storage, and predictive analytics [31]. This integration facilitates not only the optimization of renewable energy distribution based on consumption patterns but also enhances grid stability and reduces carbon emissions. Moreover, the advent of smart meters and IoT devices within these grids allows for more informed consumer choices, promoting energy savings and fostering a more environmentally friendly consumption model.

Baboli et al. [32] looks into identifying the time-varying load dynamics within the distribution grid, a complexity arising from the integration of distributed energy resources (DERs) and influenced by climate change, new consumer load patterns (e.g., electric vehicles), and renewable energy sources. Employing a combination of nonlinear numerical optimizations and system identification techniques and utilizing artificial neural networks to relate model parameters to measurement data, the study provides a novel approach to recognizing similar load dynamics without the need for extensive numerical optimization. In the case of solar PV, Jain et al. [28] introduces a DT approach for distributed PV system fault diagnostics in buildings and rooftops, focusing on real-time estimation of PV energy conversion unit outputs for detecting and identifying faults with high sensitivity. Experimental validation shows the approach's capability to detect and identify various faults in PV components swiftly, outperforming existing methods in fault sensitivity and diagnostic speed.

Fahim et al. [33] proposed a model that enhances wind farm management by using a 5G-assisted cloud-based DT framework for real-time monitoring and predictive analysis of wind speeds and power generation, showing superior performance over classical models through deep learning and K-Nearest Neighbors (K-NN) regression, based on experiments with onshore wind turbine data. Li et al. [34] introduces a DT-driven sensing methodology to address the inaccuracies and failures of wind speed sensors in wind turbines, utilizing virtual sensor arrays for fault identification and data reconstruction based on a spatiotemporal network. The methodology significantly improves sensor reliability and turbine operation safety, as demonstrated by a comprehensive analysis showing a root mean square error of 0.45 and an uncertainty of ±0.009, outperforming baseline models in engineering applications. Demonstrated on IEEE 33-node and real-world networks, Yin et al. [35] introduced a DT framework for real-time fault identification in distribution networks with integrated distributed wind power, utilizing fully electromagnetic transient calculations and skip-connected dilated causal convolution for multi-time-scale data analysis. The framework achieves ultra-real-time, high-precision fault identification, and simulation, significantly enhancing traditional post-fault analysis by accelerating it to microsecond-level situation awareness.

In the domain of hydropower, Moussa et al. [36] developed a DT of a large hydro generator using finite element simulation to design, monitor, and conduct tests, including analysis under normal operations and at the limits of its capability curve. This case study examined a three-phase synchronous machine generator rated at 310 MVA, with 56 poles and 13.8 kV. The model was verified in accordance with IEEE Standard 115, and the damper bar currents and air gap magnetic flux density distribution in both under- and over-excitation modes were analyzed. Ebrahim et al. [37] highlighted the critical importance of digital twin (DT) technology in ensuring the optimal design and reliable operation of large generators for future renewable energy systems, while also pointing out the lack of a real strategy and thorough investigation on this concept. They explored the need for and challenges associated with DT models for large-scale renewable energy producers and developed a comprehensive modeling proposal to create a multi-domain live simulation platform for hydropower and wind farms.

2.3.1.3 Microgrid Security

Cai et al. [38] proposed a comprehensive energy management framework for renewable hybrid AC-DC microgrids, incorporating various renewable resources, energy storage, and PHEVs, optimized with a bat optimization algorithm to reduce operating costs for both island and grid-connected operations. It also integrates an intrusion detection system using sequential hypothesis testing to protect wireless metering infrastructures against cyber-attacks, with the framework's reliability and efficiency validated on an IEEE 33-bus hybrid microgrid model. Danilczyk et al. [26, 39] introduced a new DT framework named *ANGEL*, aimed at enhancing the security and durability of microgrids. This framework not only enables real-time data visualization of both cyber and physical layers of the microgrid but also provides capabilities for diagnosing and healing the system, thereby mitigating component failures and the impacts of cyber attacks, with its effectiveness demonstrated through a simulation on the IEEE 39-bus benchmark. Another study [40] developed an IoT-based DT that addresses a wider spectrum of attack resiliency, such as denial of service (DoS), fault injection, etc., in order to prevent microgrid cyberattacks.

A perspective article [41] explored DT applications across various stages of microgrid development, from design to operation and maintenance, highlighting four key applications, one of which is microgrid protection. They talked about the unique problems that the microgrid protection system must consider and how DTs can offer practical solutions to these problems. Atalay et al. [42] proposed a DT-based methodology to support cybersecurity in smart grids, an essential component of industrial control systems (ICS), which face increased risks from the rapid expansion of IoT. The goal of this approach is to provide a standardized framework for extensive and ongoing penetration testing across the smart grid's lifecycle, effectively modeling its physical operations to prevent disruptions during security evaluations.

2.3.1.4 5G and Wireless Sensor Networks

While there has been rapid advancement in the development of DTs for different aspects of the smart grid, their use in 5G and 6G networks is still relatively new [43]. However, there is no denying that DTs have a huge potential to aid in the creation and implementation of the complex 5G ecosystem. Zhou et al. [44] proposed a secure and latency-aware resource allocation approach with DT assistance, along with a federated learning-based DT framework for 5G networks, focusing on accuracy, latency, and security. It optimizes DT performance, showing superior results in reducing iteration delays and, energy consumption while ensuring secure and prioritized access. Lopez et al. [45] examined the role of AI and DTs in enhancing B5G communication networks within power grids, focusing on managing dynamic data and evolving authorization policies. Their analysis emphasizes the potential for DTss to advance fault detection and cybersecurity, aiming for autonomous, self-learning systems in smart grids. Another work by Xu et al. [46] took a different approach by integrating 5G, UAVs, Beidou Positioning, artificial intelligence, and DTs to enhance the transmission grid's efficiency and safety through intelligent inspection and monitoring. Their research demonstrates how this combination enables remote centralized controls, intelligent fault diagnosis, autonomous problem tracking, and immersive monitoring operations, significantly improving the efficiency of transmission line inspections and the overall economic performance of power grid operations.

The contribution of DTs is also noteworthy in the field of energy management for wireless sensor networks. Fu et al. [47] proposed a cascaded sensor dynamic activation and information fusion algorithm that optimizes both energy efficiency and sensing performance in wireless sensor networks. By using a combined state ranking activation and event-driven mechanism approach and improved distributed Kalman information fusion, the algorithm reduces energy consumption, minimizes sensing errors, and demonstrates superior performance with nearly zero dead nodes and a 62.1% reduction in average error in simulations. Another work by Sakhri et al. [48] proposed a DT-based energy-efficient Wireless Multimedia Sensor Network (WMSN) monitoring system for wetlands, which optimizes energy use by combining local audio identification, image compression, and real-time feedback. The integration of DT technology enhances system performance by enabling dynamic adjustments, real-time decision-making, and superior energy efficiency compared to conventional WMSN systems, as validated through real-world experiments.

2.3.2 Electric Transportation

The transportation industry's rapid growth is leading to an ongoing energy crisis, resulting in a growing supply-demand gap. To solve this, DTs emerged by transforming electric transportation, including electric vehicles (EVs), aviation, and railways, by optimizing energy efficiency and system reliability (see Fig. 2.3 for an overview of DTs in electric transportation).

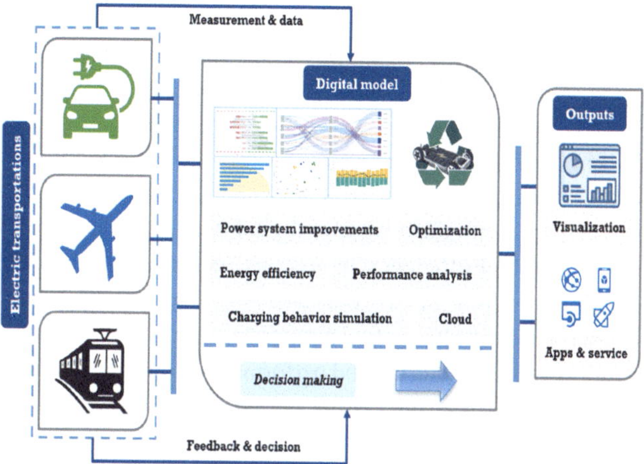

Fig. 2.3 Digital twins in electric transportation

2.3.2.1 Electric Vehicle

The integration of IoT with EVs has become the hallmark of the transportation sector, enhancing vehicle efficiency, performance, and connectivity. IoT-equipped EVs utilize real-time data analytics to optimize energy consumption and battery management, significantly extending their range and reliability. The applications of DTs in electric vehicles range from Li-ion batteries to motors and propulsion systems. We discuss energy storage devices e.g., Li-ion batteries or ultra-capacitors in Sect. 2.3-D, while here we focus on other aspects of DT applications on EVs.

The growing popularity of electric vehicles and their integration into V2G-CPSs enhances flexibility and reliability but also raises vulnerability to cyber-physical threats, particularly coordinated cyber attacks (CCAs). To address this, Ali et al. [49] proposes a resilient framework utilizing DT technologies and LSTM-based deep reinforcement learning for effective CCA detection and mitigation, demonstrating its efficacy through case studies on an IEEE 30 bus system-based V2G-CPS with rapid response times. Another work by Rassolkin et al. [50] aims to outline the development of an unsupervised prognosis and control platform for optimizing the energy systems of autonomous electric vehicles, emphasizing the creation of DTs that integrate physical entities, their virtual counterparts, and interconnected data for comprehensive performance estimation. Zhang et al. [51] took a different approach by presenting a DT of real-world electric vehicles (EVs), simulating their mobility and charging behaviors to evaluate charging algorithms and pile arrangement policies. This simulation aids in assessing the impact of dynamic traffic patterns on charging infrastructure efficiency, guiding the optimization of charging pile deployment and utilization. Ahmadi et al. [52] introduced an approach to implement the DT concept at both the equipment and system levels of Electric Railway Power Systems (ERPSs), highlighting DT's ability to accurately represent current and future states of ERPS operations. This implementation ensures reliable and secure operations by playing a crucial role in the monitoring and control functions of the complex electric railway networks.

2.3.2.2 Aviation

In aviation energy management, the advent of digital twins marks a groundbreaking shift, enabling optimization of fuel efficiency, emissions reduction, and overall sustainability through detailed virtual modeling and live data analysis. Xu et al. [53] discussed a DT-driven optimization method for 2-stroke heavy fuel aircraft engines, enabling efficient virtual simulation and optimization of manufacturing and performance. The method improved power and gas exchange performance by 4%, integrating virtual and real-world system development. A seminal work by Keller et al. [54] introduced an advanced program: "Aircraft Electrical Power Systems Prognostics and Health Management (AEPHM)", developed by Boeing, Air Force Research Laboratories, and Smiths Aerospace, focusing on diagnostics, prognostics, and decision aids for critical vehicle systems. The program, advancing through phases, leverages data from digital controllers for system health monitoring, aiming to enhance mission reliability and reduce lifecycle costs.

2.3.2.3 Multipurpose Transportation Models

DTs play key roles by providing multi-purpose, generic models that enable real-time simulation, analysis, and optimization of various transport systems, from urban mobility solutions to global logistics networks. González et al. [55] discussed the development and application of a DT for a vertical transportation system, utilizing object-oriented modeling for adaptability to various monitoring scenarios. The study demonstrated the DT's effectiveness in estimating installation parameters and simulating corrective actions through reduced models and scaled test bench measurements. Liu et al. [56] developed a maritime transportation model using DTs and IoT technologies that enhance safety performance through relay cooperation, where interference information plays a crucial role in securing data and allowing the model to harvest energy for improved communication. This energy-efficient approach not only boosts the system's transmission and security capabilities but also demonstrates optimal performance and lower outage probability through simulation, offering a promising foundation for future intelligent and secure maritime transportation.

2.3.3 Building Energy Consumption

Digital Twin significantly enhances the bottom line of smart building energy management by optimizing energy efficiency. Through real-time simulation and analysis of energy consumption patterns, DT identifies inefficiencies and proposes actionable insights for optimization [57]. This not only enhances the operation of individual buildings but also plays a crucial

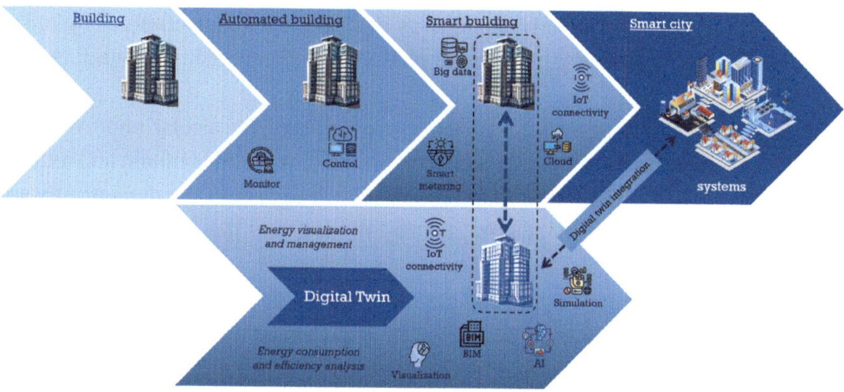

Fig. 2.4 Digital twin in building energy management

role in the transition from isolated smart buildings to interconnected smart cities. We discuss building energy management in terms of residential and commercial settings. Figure 2.4 shows an overview of DT in building energy management.

2.3.3.1 Residential

Zhou et al. [58] have come up with a cutting-edge building digital twin (BDT) framework by merging computer vision with Building Information Modeling (BIM), enabling real-time synchronization between physical buildings and their digital counterparts. It focuses on dynamic object detection and positioning through video feeds, improving the accuracy and interaction between the physical and digital realms. This method shows promise in significantly reducing location errors and enhancing building model management through real-world applications.

Khajavi et al. [59] demonstrated a methodology for creating a DT of building facades using environmental sensor networks, focusing on optimizing sensor configurations for accurate real-time modeling. It emphasizes the integration of sensors to measure parameters like light, temperature, and humidity, detailing technical challenges and solutions for establishing effective DTs. The study showcases the potential benefits of DTs in building management, energy efficiency, and architectural design, illustrating its application through a case study and discussing future research directions for indoor and larger-scale implementations.

Hadjidemetriou et al. [60] introduced a DT architecture designed for smart buildings to simulate and analyze energy efficiency, occupant comfort, and air quality in real-time and through high-granularity offline analysis. It incorporates ordinary differential equations for real-time simulation and computational fluid dynamics for detailed offline analysis. Validated through real-world experiments, this framework facilitates accurate building operation simulations, enabling effective energy management and operational optimization by integrating with a user-friendly software platform for comprehensive building analysis and control.

Song et al. [61] showcased an advanced control strategy for a solar photovoltaic (PV) system integrated with energy storage, aiming to optimize self-consumption and reduce grid dependency in residential buildings. It introduces a novel algorithm that dynamically adjusts the energy flow between the PV system, battery storage, and the grid based on real-time data and forecasts. This approach significantly enhances energy efficiency, ensures a more sustainable energy profile for homes, and demonstrates potential cost savings on electricity bills. In the case of smart homes, Huang et al. [62] presented an artificial neural network-based hour-ahead demand response (DR) algorithm for energy management that forecasts steady costs to manage price uncertainty. The algorithm's effectiveness is validated through simulations with different types of household devices, showing significant reductions in energy costs and consumer discomfort compared to systems without DR.

2.3.3.2 Commercial

Clausen et al. [63] presented a DT framework designed to enhance energy efficiency and occupant comfort in public and commercial buildings by integrating Model Predictive Control (MPC) with building automation systems. This framework utilizes weather forecasts, occupancy predictions, and current state data to optimize building operations for comfort and efficiency. Francisco et al. [64] utilized smart meter data to develop daily, temporally segmented building energy benchmarks, revealing periods of over- or underperformance that are masked by traditional annual benchmarks. These insights pave the way for DTÃƒÂ¢Ã¢âŁšÂ¬Ã¢â,¬Å“enabled urban energy management platforms, allowing for the identification of specific retrofit strategies and the achievement of near real-time efficiency improvements. Jafari et al. [65] presented a comprehensive overview of DT Technology in the realm of building operations and maintenance, emphasizing the integration of predictive modeling and control for enhanced O&M planning. It highlights the pivotal role of DTs in achieving operational efficiency, reducing maintenance costs, and improving building lifecycle management through real-time data analysis and simulation.

2.3.4 Energy Storage

A revolutionary trend toward the integration of advanced digital technology in battery energy storage systems and wider energy applications is seen in recent studies [9], [5], [12], [66] on the use of DTs in energy management. These studies demonstrate new approaches and trends as well as the vital role that DTs play in resolving industrial issues, maintaining cybersecurity, and opening the door for further developments in the energy storage system.

2.3.4.1 Batteries

Batteries are the most common form of chemical energy storage and serve as the primary energy storage systems in vehicular applications. They efficiently convert chemical energy into electrical energy to power various vehicle components, ensuring reliable performance. For electric vehicle energy storage systems, Vandana et al. [13] present a multidimensional DT that includes phases for research and development, production, operation, and recycling. This integrated strategy demonstrates an inventive breakthrough in battery lifecycle management by utilizing IoT, AI, and cloud computing to improve battery performance, cybersecurity, and longevity while addressing material shortages through recycling techniques. A DT model for Battery Energy Storage Systems (BESS) is presented by Kharlamova et al. [67] with the goal of improving frequency regulation capabilities. This approach considerably increases the accuracy of SOC (State of Charge) predictions by utilizing artificial intelligence. Their case study highlights the potential of this technique to guarantee the steady and effective operation of BESS inside contemporary energy systems and shows how effective it is in real-world scenarios. To aid in the cyber-security investigation, Shitole et al. [68] present a reasonably priced Real-Time Digital Twin (RTDT) for Residential Energy Storage Systems (RESS). Through the use of a single board computer and Simulink Desktop Real-Time, the RTDT provides an adaptable and affordable method of simulating micro-grid test beds or Cyber-Physical Systems. Through an experimental case study, this paper validates the RTDT's performance, demonstrating its potential for commercial RESS and Cloud-based Energy Management Systems (CEMS), beyond theoretical applications.

Li et al. [69] developed a cloud-based battery management system using cloud computing and IoT to diagnose batteries, specifically focusing on SoC and SoH estimation for lithium-ion and lead-acid types. By using Particle Swarm Optimization for SoH and an Adaptive Extended H-infinity Filter for SoC prediction, the system shows enhanced battery management using DTs. Validation results affirm the system's accuracy and reliability in SoC and SoH estimations, enhancing management for stationary and mobile uses. Tang et al. [70] designed a DT-based battery management system (BMS) focusing on the accurate estimation of SoC using a novel joint HIF-PF (Hybrid Information Filter-Particle Filter) algorithm. This system addresses the limitations of traditional BMS in terms of data sharing, processing capability, and storage capacity, significantly enhancing SoC estimation accuracy. Figure 2.5 shows the DT structure in energy storage for battery management systems.

2.3.4.2 Distributed Storage and Supercapacitors

Apart from batteries, several other energy storage systems (ESS) are being integrated with the Internet of Things (IoT) to enhance their efficiency, reliability, and smart management. Gao et al. [71] introduces a novel distributed energy management strategy for microgrids with high penetration of renewable energy sources and stochastic loads, addressing power system challenges by decomposing the optimization problem into two levels and considering loads as stochastic variables. Utilizing the moth flame optimization algorithm and

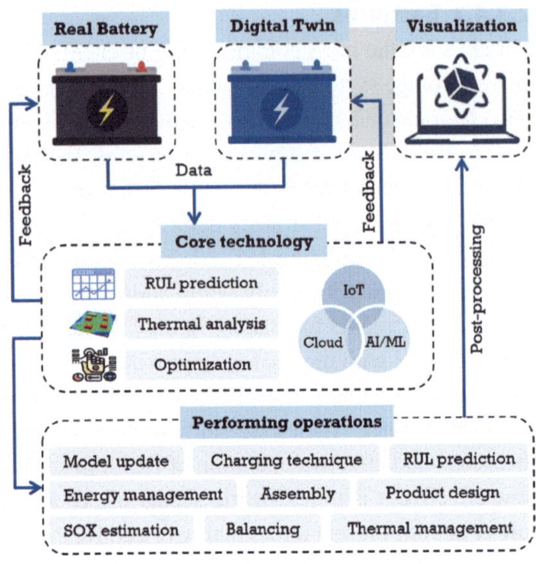

Fig. 2.5 A digital twin structure in energy storage for battery management

integrating various renewable sources and batteries, the developed distributed EMS demonstrates efficiency and applicability in practical simulations. Furthermore, several papers [72, 73] developed DTs to provide estimations and control of the charge of a supercapacitor for enhanced energy storage. This becomes especially critical given that all data on the supercapacitor are collected and transmitted to the cloud via IoT, leading to the creation of a DT model for the supercapacitor management system. These DT models can utilize real-time data within the data interaction network to assess the operational status of the supercapacitor. O'Dwyer et al. [74] introduced an energy management tool leveraging IoT-based DT technologies for optimal control, scheduling, forecasting, and coordination of energy assets, including energy storage, across urban districts to meet environmental, economic, and resilience goals. This tool, which is adaptable across various energy sectors, enables local governments to manage assets like low-carbon heating and electric vehicle charging in a coordinated manner, as demonstrated in case studies from Greenwich, London. Zhang et al. [75] enhanced photovoltaic power prediction with a DT-based framework that ensures accurate data transmission and mapping from physical to digital realms. The approach uses a Generative Adversarial Network (GAN) for restoring historical data, improving DT quality, and introducing a new prediction technique combining a digital-physical model with a CNN-BiLSTM network.

2.3.4.3 Thermal and Gas

The integration of IoT technology into thermal systems has transformed the way energy storage and heating systems operate, making them more efficient and easier to control. IoT-based frameworks are now being used in a wide range of applications, from heat storage systems in smart buildings to solar thermal power plants, helping optimize energy management [76]. Deng et al. [77] proposed a Metaverse-driven system for energy storage, including technologies like digital twins and AI for real-time monitoring, fault detection, and remote management. It uses a GA-BP neural network for accurate power load prediction, which applies to various storage types, including thermal storage, enhancing efficiency, and reducing costs. Shen et al. [78] focused on the thermal management of cell-to-pack (CTP) battery systems for electric vehicles. It develops a detailed thermal model using a liquid cooling plate structure to optimize temperature control during charging and cooling processes. The study offers insights for improving CTP battery thermal management design, applicable to various storage systems, including thermal storage.

Kang et al. [79] developed a DT model for a plant-scale solid oxide fuel cell (SOFC) system using hydrogen as the primary fuel source. It focuses on scaling the model from a 1 kW pilot plant to a 25 kW commercial plant, optimizing temperature and fuel flow rates to ensure efficient hydrogen conversion and power generation. Zhao et al. [80] came up with a DT model for high-temperature membrane electrolyzers with proton exchange using hydrogen generation. It combined multiphysics simulations and system identification to predict dynamic behaviors like power consumption, temperature, and hydrogen output, optimizing control strategies for improved system stability and efficiency. Meraghni et al. [81] introduced a data-driven DT prognostics method for proton exchange membrane fuel cells (PEMFCs) to predict their remaining useful life (RUL). The approach integrates a stacked autoencoder model with transfer learning, allowing the system to adapt to operational changes and limited data, thereby improving accuracy in hydrogen-based energy systems. Wang et al. [82] proposed a multi-physics DT for proton exchange membrane fuel cells (PEMFCs), combining 3D modeling and machine learning to predict temperature, gas, and humidity fields accurately. It enhances hydrogen-based PEMFC efficiency and real-time monitoring.

2.3.5 Others

Beyond the literature reviews categorized in various subsections discussed above, there are several other noteworthy works that we will explore here i.e., cyber-physical power systems, generic power systems, and online power analysis tools.

2.3.5.1 Cyber-Physical Power Systems

IoT-based energy systems and cyber-physical energy systems (CPES) share similarities but are not exactly the same thing [83, 84]. They differ primarily in scope and focus, though they overlap in certain areas. CPES refers to a broader, more integrated system where physical components (e.g., power plants, grids, energy storage systems) are tightly coupled with computational models and algorithms. CPES emphasizes the integration of physical and cyber layers—the energy infrastructure and digital control systems—to create feedback loops for monitoring, automation, and optimization. Due to the strong similarity with generic IoT-based systems, we highlight a few notable research works in this field.

When focusing specifically on electrical energy systems, the term Cyber-Physical Power Systems (CPPS) is also prominent. Several notable surveys have explored various aspects of CPPS, including modeling methods for future energy systems [85], smart grids [86], and applications in cybersecurity [87]. CPPS modeling is closely related to DT, as both involve the integration of physical and digital components for real-time monitoring, control, and optimization. In CPPS, modeling is used to simulate the interactions between physical power systems and their cyber layers, such as control algorithms and communication networks. DTs take this concept further by creating dynamic, real-time replicas of the physical system, continuously updated with live data. In other words, DTs appear as an advanced form of CPPS modeling that provides a more detailed and responsive representation of power systems. For instance, Yohanandhan et al. [87] discussed CPPS modeling as three primary categories: interconnection, interdependent, and interaction modeling [see Fig. 2.6]. Interconnection modeling focuses on the distinct roles of the physical and cyber systems, treating them separately. Interaction modeling, on the other hand, examines the influence each system has on the other. Interdependent modeling evaluates the extent to which the two systems

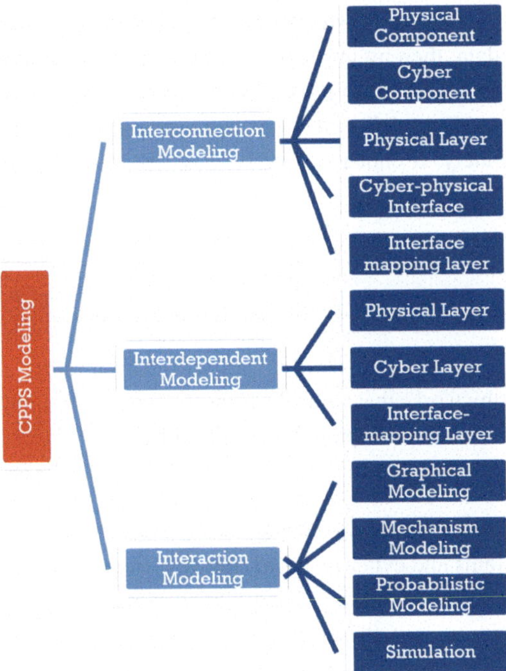

Fig. 2.6 CPPS modeling categories [87]

rely on one another. Although these categories may seem similar, each offers a unique perspective on CPPS. Interconnection modeling typically targets individual components, while interdependent modeling takes a broader, system-level approach.

2.3.5.2 Generic Power Systems

Several DT applications emphasize the development of architectural and comprehensive frameworks designed for the broad spectrum of generic power systems. Vargas et al. [88] designed a modular framework for implementing Power System Digital Twins (PSDTs), offering a flexible, reliable, and affordable approach. It highlights the significance of PSDTs in enhancing network operation, planning, and monitoring across the power industry. By adopting a modular design, the framework enables expansion beyond specific power system components, which in turn makes it easier to integrate various services and users. The implementation of this framework is demonstrated through a real-time model of the Australian National Electricity Market, showcasing its application in renewable energy integration and prospective scenario analysis. Kummerow et al. [89] structured a DT framework for enhancing the cybersecurity and operational reliability of power systems, focusing on the integration of DTs at both control center and substation levels. It discusses the benefits of DTs in improving dynamic observability, anomaly detection, and counteraction against cyber-physical threats through real-time simulations and data analysis. Additionally, the paper presents a holistic approach to secure communication and data management, showcasing a test bench for validating DT-based services in power systems. Kulikov et al. [90] developed a DT model for electricity systems to manage emergencies and operational decisions in electric power systems, using a mathematical model that merges virtual and real components for real-time simulation and response. It introduces a method to resolve system contradictions and ensure controllable, non-contradictory models, facilitating efficient emergency recovery and system functionality restoration. This approach enhances the predictability and operational efficiency of electric power systems, aligning with Industry 4.0 and digital economy transitions.

Tang et al. [91] outlined the development of multi-timescale DT models for regional multiple energy systems (RMES) on the CloudPSS platform, aimed at improving energy utilization and system optimization. It discusses constructing DTs with power flow and electromagnetic transient models to cater to varied analysis needs. Validated through the IEEE-13 and a photovoltaic test system, the models demonstrate effective system parameter verification, abnormal device detection, and optimized operation, showcasing the practicality of DTs in ensuring the safety and stability of RMES. Dufour et al. [92] designed a Hardware-In-the-Loop (HIL) testing of modern onboard power systems using DTs, focusing on their application in navy ships. It emphasizes the importance of HIL simulation in safely and effectively integrating, testing, and verifying complex power systems, reducing risks and costs associated with direct subsystem interconnection. By employing a Model-Based Design approach, the paper highlights how real-time simulators facilitate the development and integration of these systems, enabling detailed testing across various control hierarchy levels, ultimately improving system reliability and performance. Another work [93] introduces a DT that merges multi-disciplinary and multi-scale simulation processes using physical models and sensors to map the entire lifecycle of entities in a virtual space, analyzing existing DT models, their applications, and implementation methods in various fields. This technology, when applied to the power system, facilitates the optimization design of power grids, simulation of faults, management of virtual power plants, and monitoring of intelligent equipment, with the key technologies of PSDT illustrated through the Cai-Lun Station example.

2.3.5.3 Online Power Analysis

With an emphasis on real-time analysis systems to manage the complexity of contemporary power grids, Zhou et al. [94] presented a DT framework for power grid online analysis. The system uses machine learning-based security assessment, complex event processing, high-performance parallel computing, and in-memory computing to create an Online Analysis DT (OADT) that can imitate a power grid in real-time with delays of less than a second. This approach has been tested on a large-scale network model and deployed in a provincial dispatching center in China, showcasing its ability to enhance the efficiency and security of power grid operations. Liu et al. [95] explored the implementation of DT technology for online status evaluation of key power transmission and transformation equipment. It highlights the technology's capacity for enhancing real-time data accuracy, equipment status modeling, and fault prediction by integrating big data analysis and IoT. The framework consists of a layered architecture including data, platform, application, and display layers, facilitating comprehensive monitoring, diagnosis, and predictive analysis. Duan et al. [96] introduced a DT-based online tool for electric power communication system training, focusing on enhancing training and daily operations for staff and communication commissioners through a virtual-real interactive, comprehensive system. It covers the design, modeling, control principles, data collection, and visual display of DT tools, validated through a case study on optical fiber damage recovery in communication malfunctions.

2.4 Discussions

2.4.1 Current Trends in DTs in Research

We have analyzed the trend of notable research works published focusing on the usage of DTs in IoT-driven energy systems within the last few years. Figure 2.7 illustrates a clear upward trend in the number of documents published from 2017 through 2023. The starting point in 2017 shows a relatively low number of documents, which suggests that the topic might have been in its nascent stages within the scholarly or industry community at that time.

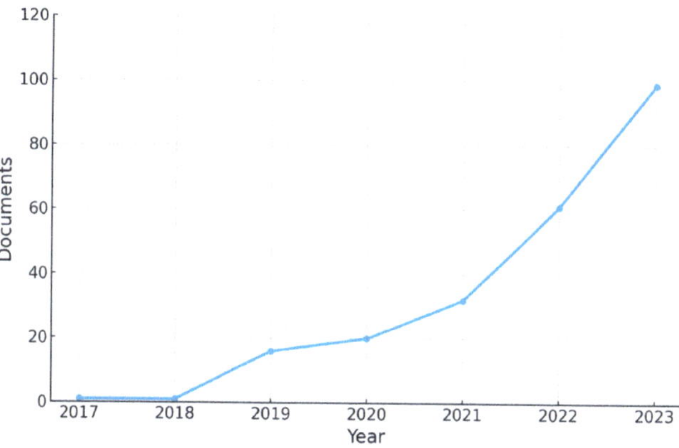

Fig. 2.7 Publications by year

There is a consistent increase in publications year-over-year, indicating growing interest and possibly advancements in the field. This is due to an increase in the technology's capabilities, wider adoption of IoT in energy systems, and more research funding being allocated to this area. A notable surge occurs between 2021 and 2023 which points to a breakthrough in the field and a realization of the practical applications of digital twins in energy systems, leading to more research and publications.

The topic is covered across various sources, indicating multidisciplinary interest and applications. Figure 2.8 shows the number of documents categorized by the source of publication from 2019 to 2023. Most sources show an increasing trend in publications over the years, with IFAC PapersOnline and Procedia CIRP showing the most significant growth, reflecting their respective communities' increasing focus on this area. The ACM series

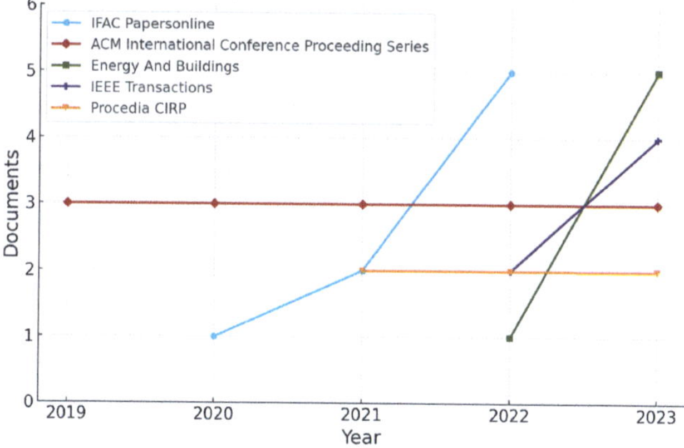

Fig. 2.8 Publications by source

shows a possible decline in 2023, due to various factors, including shifts in research focus and the specific themes of conferences for that year. IEEE Transactions has maintained a stable interest over the years, reflecting a continuous but niche focus on the topic within the engineering and technical community. The increase in energy and building sources is attributed to the rising relevance of DTs in building energy management systems and the growing need for energy efficiency in buildings.

2.4.2 Challenges in Quantifying DT Contributions

It is difficult to develop quantitative metrics to evaluate DT's contributions to energy systems due to the complexity and dynamic nature of both the systems and the DTs themselves. Energy systems involve numerous interconnected components and variables, and DTs continuously evolve by integrating real-time data, making it hard to define fixed metrics. Additionally, the varied applications of DTs across different scales—from microgrids to large power plants—complicate the creation of standardized metrics that apply universally. Data uncertainties, the interdisciplinary nature of energy optimization, and the lack of standardized frameworks further hinder the development of reliable and comprehensive quantitative metrics. Nevertheless, some studies [97, 98] have developed their own specific quantitative metrics to measure their contributions.

2.4.3 IEEE Bus Systems

The IEEE bus systems are standard models used for testing and benchmarking in power system studies, including DT applications. The adoption of DTs across various IEEE bus systems—ranging from the simpler 9-bus, 14-bus, 33-bus, and 34-bus models to the more complex 30-bus, 39-bus, and 118-bus systems—demonstrates their versatility and efficacy in simulating diverse electrical networks. These bus systems, each with unique configurations and challenges, serve as standardized benchmarks for testing and refining various DT technologies. Much of the literature has utilized bus systems as either validation testbeds or benchmarks for implementation studies. Test benchmarks serve to meet diverse research requirements and enhancements, categorized into small-scale and large-scale test cases, facilitating targeted analysis and development [99]. Table 2.2 outlines an overview of the different DT applications across various IEEE bus systems, highlighting their versatility in adapting to different scales and requirements in power system management.

Table 2.2 DT application in various IEEE bus systems

IEEE bus systems	Reference
9-bus	[26]
14-bus	[100, 101]
30-bus	[25, 49]
33-bus	[35, 38, 102–105]
34-bus	[106]
39-bus	[26, 39, 107]
118-bus	[49, 107]

2.4.4 Challenges and Future Directions

In order to achieve a level of development where EDTs can be virtually indistinguishable from their physical counterparts necessitates addressing a multitude of challenges, such as:

- Data Synchronization: Ensuring real-time, seamless synchronization between the DT and its physical counterpart to reflect the current state of the energy system accurately.
- Interoperability: Ensuring compatibility among diverse systems and technologies within the IoT ecosystem to enable comprehensive integration and communication between the DT and other systems.
- Cybersecurity: Given the DT's reliance on data and connectivity in the smart grid, safeguarding it from cyber attacks is essential to maintaining the integrity and confidentiality of sensitive information [108].
- Accuracy and Reliability: Enhancing the precision of simulations and energy consumption predictions made by the DT to ensure they are reliable and can be trusted for critical decision-making [33].
- Cost and Resource Efficiency: Addressing the high costs and substantial computational resources required to develop and maintain highly sophisticated DTs.
- User Training and Adoption: Ensuring that stakeholders are adequately trained to use and interpret the data and insights generated by the DT, which is essential for its effective adoption and use.
- Long-term Maintenance and Updating: Establishing processes for the ongoing maintenance and updating of DTs to ensure they remain accurate over the lifecycle of their physical counterparts.

After properly dealing with various challenges we can build the future beginning with this statement: *"AI-enhanced digital twins will enable autonomous energy management, integrating 100% renewable sources into smart grids, optimizing systems in real-time, and boosting sustainability through predictive maintenance, automated prototyping, and resilient cybersecurity frameworks"*. We believe that artificial intelligence and 100% renew-

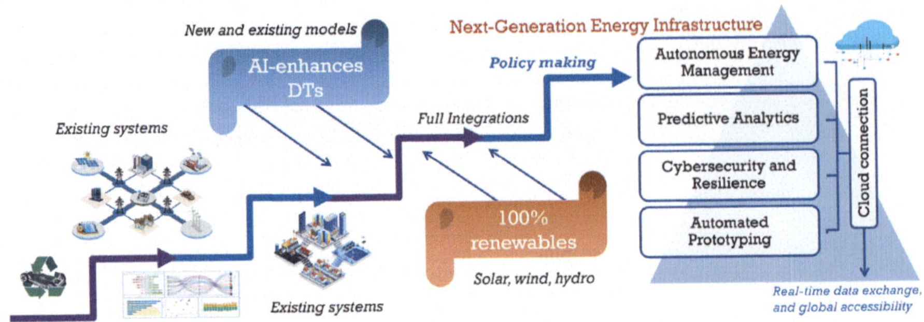

Fig. 2.9 Future roadmap for DT applications in energy

able energy resources can pave the way to the future of energy systems. By anticipating inefficiencies and possible system breakdowns, AI-powered DTs improve energy efficiency and minimize downtime, enabling optimum operations in renewable energy sources. AI is used in energy trading to process real-time data and understand intricate market trends, allowing companies to make profitable and well-informed trading decisions. To achieve this we showcase a brief roadmap (see Fig. 2.9) as follows:

Policy Making: With the integration of 100% renewable energy through digital twin technology, the development of robust and adaptive policies becomes essential. These policies must address regulatory frameworks, grid management strategies, and sustainability goals to ensure a smooth transition. Policymakers need to consider the dynamic nature of energy markets, the rapid pace of technological advancements, and the necessity for resilient infrastructure to support renewable integration.

Autonomous Energy Management: AI-driven autonomous systems allow for self-optimization of energy production and consumption. These systems make real-time decisions to maximize efficiency, reduce human intervention, and adapt dynamically to changing energy demands and renewable energy availability. This reduces the need for continuous involvement of individuals, as the systems are autonomous and handle operations efficiently. While some may argue that this could lead to a reduction in employment, we contend that engineers and designers will still be crucial in overseeing and managing the integration of DT and AI technologies, ensuring their proper function and continuous improvement.

Predictive Analytics: By using AI and data from DTs, predictive analytics can forecast potential system failures or inefficiencies before they happen. This allows for proactive maintenance and optimization of energy flows, increasing reliability and reducing downtime in renewable energy systems.

Cybersecurity and Resilience: As previously mentioned, cybersecurity remains one of the most critical challenges in the development of smart energy systems. As these systems become increasingly interconnected and intelligent, the risk of cyberattacks becomes more imminent. AI-enhanced DTs play a vital role in strengthening the cybersecurity and resilience of energy systems by detecting and mitigating potential threats in real-time. With renewable energy sources integrated and DTs overseeing system performance, resilient solutions can be developed to proactively defend against vulnerabilities, ensuring stability and security across the energy grid.

Automated Prototyping: DTs, when paired with AI, revolutionize the future of energy by enabling rapid, automated prototyping through virtual simulations. This allows engineers to experiment with designs, test various scenarios, and optimize systems without the time, cost, or risk of real-world trials. Automated prototyping not only accelerates innovation but also enhances precision, as AI can sift through countless variables and configurations to find the most efficient solutions. It empowers energy companies to adapt swiftly to changes, integrate renewable sources more seamlessly, and develop solutions that are both innovative and resilient—transforming ideas into reality faster than ever before. For industry practitioners, this leads to faster deployment of new technologies, minimizing time-to-market, and offering a safer environment for testing disruptive innovations without risking existing infrastructure.

Cloud Connection: The cloud unlocks real-time data exchange, acting as the backbone of scalability and global connectivity for energy systems. It delivers the immense computational power needed to process vast streams of data, enabling predictive analytics, real-time monitoring, and flawless communication across distributed energy sources. By seamlessly linking renewable power systems and smart grids worldwide, the cloud paves the way for smarter, faster, and more efficient energy management, transforming how we harness and distribute energy on a global scale.

2.4.5 Economic Integration

DTs integrate economic considerations into energy systems by optimizing performance, reducing costs, and improving overall operational efficiency. Through real-time monitoring, predictive maintenance, and fault detection, DTs minimize downtime, prevent equipment failures, and extend the lifespan of energy assets, leading to significant cost reductions. They also optimize energy efficiency by identifying and addressing inefficiencies in smart grids, building energy management, and energy storage systems, ultimately lowering energy consumption and operational expenses. For instance, Xu et al. [109] demonstrated the cost-effective impact of a DT on a 320 MW coal-fired thermal power plant, highlighting substantial improvements in operating economics through its digital model.

Additionally, DTs help evaluate the integration of renewable energy sources, balancing sustainability with cost-effectiveness. By simulating various energy mixes, DTs allow decision-makers to optimize the economic and environmental impact of energy systems. They also support lifecycle cost analysis, ensuring informed decisions that consider long-term costs such as maintenance and upgrades. In energy trading and demand-response markets, DTs improve profitability by simulating market trends and optimizing participation strategies. Furthermore, DTs can assist in investment decision-making, enabling stakeholders to explore different infrastructure and energy technology investments for maximum financial returns.

2.5 Conclusion

The emergence of the Internet of Things has catalyzed an interdisciplinary fusion, integrating artificial intelligence, computing system design, and smart industrial infrastructure, paving the way for the evolution of virtual prototyping solutions such as digital twins. These innovations, integral to IoT applications, have profoundly influenced the development and optimization of IoT-driven smart energy systems. Digital twins, by mirroring the complexities of these systems, offer useful platforms for simulation, analysis, and real-time decision-making, thereby enhancing operational efficiency, resilience, and sustainability. In this chapter, we narrowed our focus to the application of digital twins in emerging smart energy systems powered by IoT infrastructure for several compelling reasons. To begin with, the combination of IoT technology and DTs offers a novel way to improve the sustainability, dependability, and efficiency of smart energy systems. DTs can facilitate more informed decision-making and proactive maintenance strategies by utilizing real-time data from IoT devices to generate detailed simulations and predictive analytics. Furthermore, as the demand for cleaner and more efficient energy solutions grows, the need for sophisticated management and optimization tools becomes increasingly critical.

In order to address these issues, DTs provide a strong platform that makes it possible to dynamically model and optimize the production, distribution, and consumption of energy. Finally, by concentrating on this intersection, we can investigate novel solutions to challenging energy issues, advancing the energy sector's transition to more resilient and sustainable infrastructure.

References

1. Ahmad, Fahad Faraz, Chaouki Ghenai, and Maamar Bettayeb. 2021. Maximum power point tracking and photovoltaic energy harvesting for internet of things: A comprehensive review. *Sustainable Energy Technologies and Assessments* 47: 101430.
2. Digital Twins for Energy Systems. 2023. https://innovate.ieee.org/innovation-spotlight/digital-twins-for-energy-systems/. Accessed 01 Feb 2024.
3. Justin, Tuttle. 2022. *Digital twin: Driving innovation in the energy sector*. https://blogs.sw.siemens.com/energy-utilities/2022/03/23/digital-twin-driving-innovation-in-the-energy-sector/. Accessed 01 Feb 2024.
4. Yu, Wei, et al. 2022. Energy digital twin technology for industrial energy management: Classification, challenges and future. *Renewable and Sustainable Energy Reviews* 161: 112407.
5. Amaral, João, et al. 2023. Energy digital twin applications: A review. *Renewable and Sustainable Energy Reviews* 188: 113891. https://doi.org/10.1016/j.rser.2023.113891.

6. Pandiyan, Pitchai, et al. 2023. Technological advancements toward smart energy management in smart cities. *Energy Reports* 10: 648–677. https://doi.org/10.1016/j.egyr.2023.07.021.
7. Lei, Lei, et al. 2020. Dynamic energy dispatch based on deep reinforcement learning in IoT-driven smart isolated microgrids. *IEEE Internet of Things Journal* 8 (10): 7938–7953.
8. Hui, Chu Xiao, et al. 2023. Greening smart cities: An investigation of the integration of urban natural resources and smart city technologies for promoting environmental sustainability. *Sustainable Cities and Society* 99: 104985.
9. Ghenai, Chaouki, et al. 2022. Recent trends of digital twin technologies in the energy sector: A comprehensive review. *Sustainable Energy Technologies and Assessments* 54: 102837. https://doi.org/10.1016/j.seta.2022.102837.
10. Wang, Yujie, Xu Kang, and Zonghai Chen. 2022. A survey of digital twin techniques in smart manufacturing and management of energy applications. *Green Energy and Intelligent Transportation* 1 (2): 100014.
11. Lamagna, Mario, et al. 2021. A comprehensive review on digital twins for smart energy management system. *International Journal of Energy Production and Management* 6 (4): 323–334.
12. Semeraro, Concetta, et al. 2023. Digital twin application in energy storage: Trends and challenges. *Journal of Energy Storage* 58: 106347.
13. Vandana, Jagdish, Akhil Garg, and Bijaya Panigrahi. 2021. Multi-dimensional digital twin of energy storage system for electric vehicles: A brief review. *Energy Storage* 3. https://doi.org/10.1002/est2.242.
14. Chen, Haoyu, et al. 2022. Digital twin techniques for power electronics-based energy conversion systems: A survey of concepts, application scenarios, future challenges, and trends. *IEEE Industrial Electronics Magazine*.
15. Jafari, Mina, et al. 2023. A review on digital twin technology in smart grid, transportation system and smart city: Challenges and future. *IEEE Access*.
16. Bhatti, Ghanishtha, Harshit Mohan, and R. Raja Singh. 2021. Towards the future of smart electric vehicles: Digital twin technology. *Renewable and Sustainable Energy Reviews* 141: 110801.
17. Yassin, Mohammed AM, Ashish Shrestha, and Suhaila Rabie. 2023. Digital twin in power system research and development: Principle, scope, and challenges. *Energy Reviews* 100039.
18. Sifat, Md Mhamud Hussen, et al. 2023. Towards electric digital twin grid: Technology and framework review. *Energy and AI* 11: 100213.
19. Kabir, Md Rafiul, Dipal Halder, and Sandip Ray. 2024. Digital twins for Iot-driven energy systems: A survey. *IEEE Access*.
20. Cespedes-Cubides, Andres Sebastian, and Muhyiddine Jradi. 2024. A review of building digital twins to improve energy efficiency in the building operational stage. *Energy Informatics* 7 (1): 11.
21. Priyadarshini, Ishaani, et al. 2021. Identifying cyber insecurities in trustworthy space and energy sector for smart grids. *Computers & Electrical Engineering* 93: 107204.
22. Wei, Lou, Chen Yi, and Jin Yun. 2021. Energy drive and management of smart grids with high penetration of renewable sources of wind unit and solar panel. *International Journal of Electrical Power & Energy Systems* 129: 106846.
23. Khalil, Muhammad Ibrahim, et al. 2021. Hybrid smart grid with sustainable energy efficient resources for smart cities. *Sustainable Energy Technologies and Assessments* 46: 101211.
24. Moslehi, Khosrow, and Ranjit Kumar. 2010. A reliability perspective of the smart grid. *IEEE Transactions on Smart Grid* 1 (1): 57–64.
25. Tzanis, Nikolaos, et al. 2020. A hybrid cyber physical digital twin approach for smart grid fault prediction. In *2020 IEEE conference on industrial cyberphysical systems (ICPS)*, vol. 1, 393–397. IEEE.

26. Danilczyk, William, Yan Lindsay Sun, and Haibo He, 2021. Smart grid anomaly detection using a deep learning digital twin. In *2020 52nd North American power symposium (NAPS)*, 1–6. IEEE.
27. Olatunji, Obafemi O., et al. 2021. Overview of digital twin technology in wind turbine fault diagnosis and condition monitoring. In *2021 IEEE 12th international conference on mechanical and intelligent manufacturing technologies (ICMIMT)*, 201–207. IEEE.
28. Jain, Palak, et al. 2019. A digital twin approach for fault diagnosis in distributed photovoltaic systems. *IEEE Transactions on Power Electronics* 35 (1): 940–956.
29. Castellani, Andrea, Sebastian Schmitt, and Stefano Squartini. 2020. Real-world anomaly detection by using digital twin systems and weakly supervised learning. *IEEE Transactions on Industrial Informatics* 17 (7): 4733–4742.
30. Hossen, Tareq, Dushyant Sharma, and Behrooz Mirafzal. 2021. Smart inverter twin model for anomaly detection. In *2021 IEEE 22nd workshop on control and modelling of power electronics (COMPEL)*, 1–6. IEEE.
31. Javid, Iqra, et al. 2021. Futuristic decentralized clean energy networks in view of inclusive-economic growth and sustainable society. *Journal of Cleaner Production* 309: 127304.
32. Baboli, Payam Teimourzadeh, Davood Babazadeh, and Darshana Ruwan Kumara, and Bowatte. 2020. Measurement-based modeling of smart grid dynamics: A digital twin approach. In *2020 10th smart grid conference (SGC)*, 1–6. IEEE.
33. Fahim, Muhammad, et al. 2022. Machine learning-based digital twin for predictive modeling in wind turbines. *IEEE Access* 10: 14184–14194.
34. Li, Yang, and Xiaojun Shen. 2021. A novel wind speed-sensing methodology for wind turbines based on digital twin technology. *IEEE Transactions on Instrumentation and Measurement* 71: 1–13.
35. Yin, Yikun, et al. 2024. Digital twin-driven identification of fault situation in distribution networks connected to distributed wind power. *International Journal of Electrical Power & Energy Systems* 155: 109415.
36. Moussa, Cynthia, et al. 2018. Insights into digital twin based on finite element simulation of a large hydro generator. In *IECON 2018-44th annual conference of the IEEE industrial electronics society*, 553–558. IEEE.
37. Ebrahimi, Amir. 2019. Challenges of developing a digital twin model of renewable energy generators. In *2019 IEEE 28th international symposium on industrial electronics (ISIE)*, 1059–1066. IEEE.
38. Cao, Hanhua, Dongming Zhang, and Shujuan Yi. 2023. Real-time machine learning-based fault detection, classification, and locating in large scale solar energy-based systems: Digital twin simulation. *Solar Energy* 251: 77–85.
39. Danilczyk, William, Yan Sun, and Haibo He. 2019. ANGEL: An intelligent digital twin framework for microgrid security. In *2019 North American power symposium (NAPS)*, 1–6. IEEE.
40. Saad, Ahmed, et al. 2020. On the implementation of IoT-based digital twin for networked microgrids resiliency against cyber attacks. *IEEE Transactions on Smart Grid* 11 (6): 5138–5150.
41. Wu, Ying, et al. 2024. Digital twins for microgrids: Opening a new dimension in the power system. *IEEE Power and Energy Magazine* 22 (1): 35–42.
42. Atalay, Manolya, and Pelin Angin. 2020. A digital twins approach to smart grid security testing and standardization. In *2020 IEEE international workshop on metrology for industry 4.0 & IoT*, 435–440. IEEE.
43. Nguyen, Huan X., et al. 2021. Digital twin for 5G and beyond. *IEEE Communications Magazine* 59 (2): 10–15.

44. Zhou, Zhenyu, et al. 2021. Secure and latency-aware digital twin assisted resource scheduling for 5G edge computing-empowered distribution grids. *IEEE Transactions on Industrial Informatics* 18 (7): 4933–4943.
45. Lopez, Javier, Juan E. Rubio, and Cristina Alcaraz. 2021. Digital twins for intelligent authorization in the B5G-enabled smart grid. *IEEE Wireless Communications* 28 (2): 48–55.
46. Xu, Ming. 2021. Research on 5G UAV smart grid inspection technology based on digital twin. In *2021 IEEE 7th International conference on cloud computing and intelligent systems (CCIS)*, 367–370. IEEE.
47. Fu, Juteng, et al. 2024. Distributed energy-efficient wireless sensing and information fusion via event-driven and state-rank activation. *Wireless Networks* 1–15.
48. Sakhri, Aya, et al. 2024. A digital twin-based energy-efficient wireless multimedia sensor network for waterbirds monitoring. *Future Generation Computer Systems* 155: 146–163.
49. Ali, Mansoor, et al. 2023. A smart digital twin enabled security framework for vehicle-to-grid cyber-physical systems. *IEEE Transactions on Information Forensics and Security*.
50. Rassõlkin, Anton, et al. 2019. Digital twin for propulsion drive of autonomous electric vehicle. In *2019 IEEE 60th international scientific conference on power and electrical engineering of Riga technical university (RTUCON)*, 1–4. IEEE.
51. Zhang, Tianle, et al. 2019. Time series behavior modeling with digital twin for Internet of Vehicles. *EURASIP Journal on Wireless Communications and Networking* 2019: 1–11.
52. Ahmadi, Miad, et al. 2021. Adapting digital twin technology in electric railway power systems. In *2021 12th power electronics, drive systems, and technologies conference (PEDSTC)*, 1–6. IEEE.
53. Wenjun, Xu., et al. 2021. Digital twin-based industrial cloud robotics: Framework, control approach and implementation. *Journal of Manufacturing Systems* 58: 196–209.
54. Keller, Kirby, et al. 2006. Aircraft electrical power systems prognostics and health management. In *2006 IEEE aerospace conference*, 12. IEEE.
55. González, Mikel, et al. 2020. A digital twin for operational evaluation of vertical transportation systems. *IEEE Access* 8: 114389–114400.
56. Liu, Jun, et al. 2021. Security in IoT-enabled digital twins of maritime transportation systems. *IEEE Transactions on Intelligent Transportation Systems*.
57. Wang, Weixi, et al. 2022. Deep learning for assessment of environmental satisfaction using BIM big data in energy efficient building digital twins. *Sustainable Energy Technologies and Assessments* 50: 101897.
58. Zhou, Xiaoping, et al. 2023. Computer vision enabled building digital twin using building information model. *IEEE Transactions on Industrial Informatics* 19 (3): 2684–2692. https://doi.org/10.1109/TII.2022.3190366.
59. Khajavi, Siavash H., et al. 2019. Digital twin: Vision, benefits, boundaries, and creation for buildings. *IEEE Access* 7: 147406–147419. https://doi.org/10.1109/ACCESS.2019.2946515.
60. Hadjidemetriou, Lenos, et al. 2023. A digital twin architecture for real-time and offline high granularity analysis in smart buildings. *Sustainable Cities and Society* 98: 104795.
61. Song, Yuguang, et al. 2023. A data-model fusion dispatch strategy for the building energy flexibility based on the digital twin. *Applied Energy* 332: 120496. ISSN: 0306-2619.
62. Huang, Jueru, Dmitry D. Koroteev, and Marina Rynkovskaya. 2023. Machine learning-based demand response in PV-based smart home considering energy management in digital twin. *Solar Energy* 252: 8–19.
63. Clausen, Anders et al. 2021. A digital twin framework for improving energy efficiency and occupant comfort in public and commercial buildings. *Energy Informatics* 4 (2): 40. https://doi.org/10.1186/s42162-021-00153-9.

64. Francisco, Abigail, Neda Mohammadi, and John E. Taylor. 2020. Smart city digital twin—enabled energy management: Toward real-time urban building energy benchmarking. *Journal of Management in Engineering* 36 (2): 04019045.
65. Jafari, Mohsen A., et al. 2020. Improving building energy footprint and asset performance using digital twin technology. *IFAC-PapersOnLine* 53 (3): 386–391.
66. Semeraro, Concetta, et al. 2023. Digital twin in battery energy storage systems: Trends and gaps detection through association rule mining. *Energy* 273: 127086.
67. Kharlamova, Nina, Chresten Traholt, and Seyedmostafa Hashemi. 2024. A digital twin of battery energy storage systems providing frequency regulation. In *2022 IEEE international systems conference (SysCon)*, 1–7. Montreal, QC, Canada: IEEE. ISBN: 978-1-66543-992-3. https://doi.org/10.1109/SysCon53536.2022.9773919. https://ieeexplore.ieee.org/document/9773919/.
68. Shitole, Amardeep B., et al. 2021. Real-time digital twin of residential energy storage system for cyber-security study. In *2021 IEEE 2nd international conference on smart technologies for power, energy and control (STPEC)*, 1–6. Bilaspur, Chhattisgarh, India: IEEE. ISBN: 978-1-66544-319-7. https://doi.org/10.1109/STPEC52385.2021.9718616. https://ieeexplore.ieee.org/document/9718616/.
69. Li, Weihan, et al. 2020. Digital twin for battery systems: Cloud battery management system with online state-of-charge and state-of-health estimation. *Journal of Energy Storage* 30: 101557. ISSN: 2352-152X. https://doi.org/10.1016/j.est.2020.101557.
70. Tang, Hao, et al. 2022. Design of power lithium battery management system based on digital twin. *Journal of Energy Storage* 47: 103679. ISSN: 2352-152X. https://doi.org/10.1016/j.est.2021.103679.
71. Gao, Jiaxi, and Haiyan Huang. 2023. Stochastic optimization for energy economics and renewable sources management: A case study of solar energy in digital twin. *Solar Energy* 262: 111865.
72. Savenko, Oleg, et al. 2021. Automated control method for charging supercapacitors based on relaxation characteristics. In *2021 11th IEEE international conference on intelligent data acquisition and advanced computing systems: Technology and applications (IDAACS)*, vol. 1, 377–382. IEEE.
73. Yang, Yingze, et al. 2021. Supercapacitor digital twin management system based on cloud environment. In *2021 IEEE 23rd int conf on high performance computing & communications; 7th int conf on data science & systems; 19th int conf on smart city; 7th int conf on dependability in sensor, cloud & big data systems & application (HPCC/DSS/SmartCity/DependSys)*, 1014–1021. IEEE.
74. O'Dwyer, Edward, et al. 2020. Integration of an energy management tool and digital twin for coordination and control of multi-vector smart energy systems. *Sustainable Cities and Society* 62: 102412. ISSN: 2210-6707. https://doi.org/10.1016/j.scs.2020.102412.
75. Zhang, Xiaoyu, et al. 2023. Digital twin empowered PV power prediction. *Journal of Modern Power Systems and Clean Energy* 1–13. https://doi.org/10.35833/MPCE.2023.000351.
76. Tian, Feng. 2024. Internet of things temperature control of indirect dual tank heat storage system in solar photo-thermal power plant. *Thermal Science* 28: 1477–1484.
77. Deng, Yimin, Zhoubo Weng, and Tianlong Zhang. 2022. Metaverse-driven remote management solution for scene-based energy storage power stations. *Evolutionary Intelligence* 16. https://doi.org/10.1007/s12065-022-00769-0.
78. Shen, Kai, et al. 2022. A comprehensive analysis and experimental investigation for the thermal management of cell-to-pack battery system. *Applied Thermal Engineering* 211: 118422.
79. Kang, Jia-Lin., et al. 2021. Digital twin model and dynamic operation for a plant-scale solid oxide fuel cell system. *Journal of the Taiwan Institute of Chemical Engineers* 118: 60–67.

80. Zhao, Dongqi, et al. 2022. A data-driven digital-twin model and control of high temperature proton exchange membrane electrolyzer cells. *International Journal of Hydrogen Energy* 47 (14): 8687–8699.
81. Meraghni, Safa, et al. 2021. A data-driven digital-twin prognostics method for proton exchange membrane fuel cell remaining useful life prediction. *International Journal of Hydrogen Energy* 46 (2): 2555–2564.
82. Wang, Bowen, et al. 2020. Multi-physics-resolved digital twin of proton exchange membrane fuel cells with a data-driven surrogate model. *Energy and AI* 1: 100004.
83. Vanderbilt University School of Engineering. 2023. *What is the difference between CPS and IoT?* https://blog.engineering.vanderbilt.edu/what-is-the-difference-between-cps-and-iot. Accessed 06 Oct 2024.
84. Virmani, Charu, and Anuradha Pillai. 2021. Internet of things and cyber physical systems: An insight. *Recent Advances in Intelligent Systems and Smart Applications* 379–401.
85. Abdelmalak, Michael, Venkatesh Venkataramanan, and Richard Macwan. 2022. A survey of cyber-physical power system modeling methods for future energy systems. *IEEE Access* 10: 99875–99896.
86. Macana, Carlos Andrés, Nicanor Quijano, and Eduardo Mojica-Nava. 2011. A survey on cyber physical energy systems and their applications on smart grids. In *2011 IEEE PES conference on innovative smart grid technologies Latin America (ISGT LA)*, 1–7. IEEE.
87. Yohanandhan, Rajaa Vikhram, et al. 2020. Cyber-physical power system (CPPS): A review on modeling, simulation, and analysis with cyber security applications. *IEEE Access* 8: 151019–151064.
88. Arraño-Vargas, Felipe, and Georgios Konstantinou. 2023. Modular design and real-time simulators toward power system digital twins implementation. *IEEE Transactions on Industrial Informatics* 19 (1): 52–61. https://doi.org/10.1109/TII.2022.3178713.
89. Kummerow, André, et al. 2020. Digital-twin based services for advanced monitoring and control of future power systems. In *2020 IEEE power & energy society general meeting (PESGM)*, 1–5. https://doi.org/10.1109/PESGM41954.2020.9354468.
90. Kulikov, Gennady, et al. 2021. A digital twin model for electricity systems. In *2021 international conference on electrotechnical complexes and systems (ICOECS)*, 239–244. https://doi.org/10.1109/ICOECS52783.2021.9657362.
91. Tang, Xueyong, et al. 2020. Creating multi-timescale digital twin models for regional multiple energy systems on CloudPSS. In *2020 IEEE sustainable power and energy conference (iSPEC)*, 1412–1418.https://doi.org/10.1109/iSPEC50848.2020.9351175.
92. Dufour, Christian, Zareh Soghomonian, and Wei Li. 2018. Hardware-in-the-loop testing of modern on-board power systems using digital twins. In *2018 international symposium on power electronics, electrical drives, automation and motion (SPEEDAM)*, 118–123. https://doi.org/10.1109/SPEEDAM.2018.8445302.
93. Pan, Huaming, et al. 2020. Digital twin and its application in power system. *2020 5th international conference on power and renewable energy (ICPRE)*, 21–26. IEEE.
94. Zhou, Mike, Jianfeng Yan, and Donghao Feng. 2019. Digital twin framework and its application to power grid online analysis. *CSEE Journal of Power and Energy Systems* 5 (3): 391–398.
95. Liu, Jiaxin, et al. 2021. Research on online status evaluation technology for main equipment of power transmission and transformation based on digital twin. In *2021 IEEE 5th conference on energy internet and energy system integration (EI2)*, 3368–3373. https://doi.org/10.1109/EI252483.2021.9713501.
96. Shiqi, Guan, et al. 2023. Digital twin-based online tools for electric power communication system training: Online digital twin power communication system. *IET Generation, Transmission & Distribution* 17 (1): 146–160. https://doi.org/10.1049/gtd2.12670. eprint: https://ietresearch.onlinelibrary.wiley.com/doi/pdf/10.1049/gtd2.12670.

97. Bayer, Daniel, and Marco Pruckner. 2023. A digital twin of a local energy system based on real smart meter data. *Energy Informatics* 6 (1): 8.
98. Jamil, Harun, et al. 2024. Digital twin-driven architecture for AIoT-based energy service provision and optimal energy trading between smart nanogrids. *Energy and Buildings* 319: 114463.
99. Dolatabadi, Sarineh Hacopian, et al. 2020. An enhanced IEEE 33 bus benchmark test system for distribution system studies. *IEEE Transactions on Power Systems* 36 (3): 2565–2572.
100. Xinya, Song, et al. 2020. Application of digital twin assistant-system in state estimation for inverter dominated grid. In *2020 55th international universities power engineering conference (UPEC)*, 1–6. IEEE.
101. Landen, Matthew, et al. 2022. DRAGON: Deep reinforcement learning for autonomous grid operation and attack detection. In *Proceedings of the 38th annual computer security applications conference*, 13–27.
102. Li, Qinghui, et al. 2023. Renewable-based microgrids' energy management using smart deep learning techniques: Realistic digital twin case. *Solar Energy* 250: 128–138.
103. Chen, Jian, et al. 2022. Intelligent coordinated control strategy of distributed photovoltaic power generation cluster based on digital twin technology. In *2022 4th international conference on electrical engineering and control technologies (CEECT)*, 236–240. IEEE.
104. Shen, Zhongjie, et al. 2023. Digital twin application for attach detection and mitigation of PV-based smart systems using fast and accurate hybrid machine learning algorithm. *Solar Energy* 250: 377–387.
105. Cao, Wanlin, and Lei Zhou. 2024. Resilient microgrid modeling in digital twin considering demand response and landscape design of renewable energy. *Sustainable Energy Technologies and Assessments* 64: 103628.
106. Hong, Ying-Yi, and Gerard Francesco DG Apolinario. 2022. Ancillary services and risk assessment of networked microgrids using digital twin. *IEEE Transactions on Power Systems*.
107. Cao, Shiqi, Venkata Dinavahi, and Ning Lin. 2022. Machine learning based transient stability emulation and dynamic system equivalencing of large-scale AC-DC grids for faster-than-real-time digital twin. *IEEE Access* 10: 112975–112988.
108. Olivares-Rojas, Juan C., et al. 2021. Towards cybersecurity of the smart grid using digital twins. *IEEE Internet Computing* 26 (3):52–57.
109. Xu, Bin, et al. 2019. A case study of digital-twin-modelling analysis on power-plant-performance optimizations. *Clean Energy* 3 (3): 227–234.

Digital Twins and Modern Prototyping Technologies for Vehicular Systems

3

Digital twins and prototyping approaches for next-generation automotive, aviation, and maritime systems.

3.1 Introduction

Vehicles are increasingly becoming more central in our lives due to advancements in both the economy and technology. Yet, making vehicles smarter and more environmentally friendly introduces challenges, such as increased complexity and costs for design and maintenance. The modern vehicular system spans a wide array of domains, including automotive, aviation, and maritime industries, all of which are critical to the global economy and transportation infrastructure. After smartphones and computers, cars have become the third-largest networked device thanks to the growth and popularization of sensor technologies and the Internet of Things (IoT). This technological expansion is not only revolutionizing how vehicles are designed but also how they interact with users and their surroundings. The incorporation of digital twin technology alongside IoT advances intelligent manufacturing, enabling greater automation and precision in vehicle production. Digital twins, virtual representations of physical assets, are becoming instrumental across various sectors, including automotive, aviation, and maritime industries. In vehicles, these digital counterparts provide real-time monitoring of performance and maintenance needs, thereby optimizing the entire lifecycle of design, manufacturing, and operational use. Compared to cars, aircraft are even more sophisticated, integrating a greater volume of software and hardware components. These systems require advanced avionics for navigation, communication, and flight control-far more complex than automotive electronics. The use of digital twins in aviation allows for continuous simulation of aircraft subsystems, including propulsion, aerodynamics, and safety mechanisms, ensuring optimal performance and compliance with stringent

M. R. Kabir and S. Ray, *Digital Twins for Distributed IoT Applications*, Synthesis Lectures on Engineering, Science, and Technology, https://doi.org/10.1007/978-3-032-18938-7_3

regulatory standards. Likewise, in maritime industries, digital twins of ships and their subsystems enable real-time tracking of vessel operations, structural health, and engine performance. This allows for simulations of various navigational and operational scenarios, improving decision-making and operational efficiency.

The increasing complexity of technologically-equipped vehicles poses significant challenges for traditional development paradigms, maintenance procedures, and innovation strategies. As vehicles evolve into data-rich platforms, the entire lifecycle—from design and manufacturing to maintenance and operation-becomes a data-intensive process [1]. The introduction of digital twins into this process provides a way to manage this data effectively, offering deep insights into vehicle performance and maintenance. By simulating real-time operations and providing continuous monitoring, which reduces downtime and improves safety [2]. The ability to harness this data through technologies such as machine learning, artificial intelligence, and big data further enhances automotive manufacturing processes, leading to more efficient production, personalized transportation solutions, and improved vehicle longevity [3]. Adopting these innovative methodologies to tackle the complexities of modern vehicles necessitates a paradigm shift in automotive development, testing, and maintenance. Digital twins and various prototypes are at the forefront of this shift, bridging the gap between physical and digital systems. In the automotive sector, each vehicle can be paired with its own digital twin, creating a dynamic and real-time replica of its state and behavior. Similarly, in aviation and maritime industries, digital twins enable real-time simulations and monitoring of various components, ensuring optimal performance and efficiency. By continuously analyzing and adapting to data from physical vehicles, these digital models facilitate improved design, enhanced safety measures, and more precise maintenance schedules, thereby driving innovation and sustainability across all transportation sectors. As vehicles increasingly resemble complex consumer electronics platforms driven by software, sensors, and connectivity, ensuring automotive functional safety-defined as preventing unacceptable risks of physical harm arising from electronic and software malfunctions-has become a critical and non-trivial challenge [4].

While advanced prototyping approaches and requirements are important, a more practical understanding can be gained by reviewing established research, particularly in the context of the future of vehicles. To better understand the scope of prototyping in the vehicular industry, we provide an overview of significant ongoing research in the automotive, aviation, and maritime domains. Figure 3.1 shows a taxonomy of prototyping technologies that we discuss in this chapter.

This chapter is not limited to digital twins but also considers various other prototyping technologies, some of which may seem more abstract but are closely aligned with the concept of digital twins. Furthermore, we explore the enabling technologies behind these approaches and provide insights into future research directions that could shape the development of intelligent vehicles. In light of the literature gap and the need for a unified perspective on prototyping in automotive, aviation, and maritime domains, our review addresses the following questions:

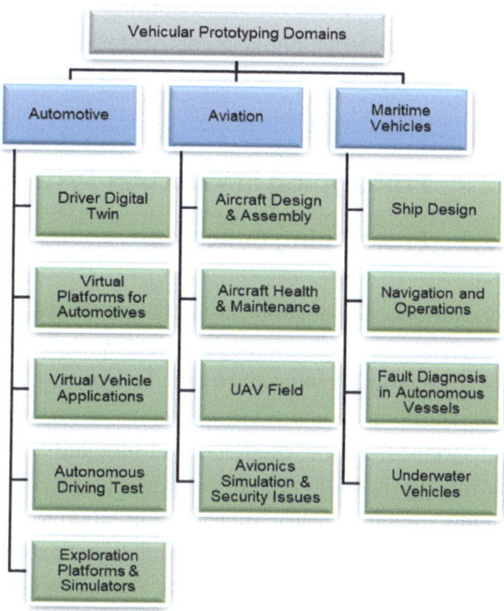

Fig. 3.1 A taxonomy of prototyping technologies of vehicular systems

- **RQ1**: Which prototyping approaches, including but not limited to digital twins, are currently being utilized across these three sectors, and how do they complement or diverge from one another?
- **RQ2**: What enabling technologies are most critical to implementing these approaches effectively, and how do they shape the advancement of intelligent vehicles?
- **RQ3**: How can insights gained from these approaches inform future research directions, promoting more effective and sustainable development of next-generation vehicular infrastructure?

These research questions are critical because they address the growing need for a comprehensive understanding of prototyping approaches across the automotive, aviation, and maritime sectors, which are increasingly converging under the broader umbrella of intelligent transportation. By identifying the attributes of existing prototyping techniques (RQ1), we can uncover gaps and opportunities for cross-sector innovations that enhance efficiency, safety, and reliability. Investigating the enabling technologies (RQ2) is essential for understanding the foundational tools and methodologies that drive advancements in smart vehicular systems, ensuring their scalability and real-world applicability. Lastly, by exploring future research directions (RQ3), we can provide valuable insights that guide policymakers, researchers, and industry leaders in developing more integrated, sustainable, and forward-looking transportation solutions. Addressing these questions will not only advance scientific knowledge but also contribute to the technological evolution of next-generation vehicular systems.

3.2 Prototyping in Vehicular Systems

3.2.1 Prototyping Concepts

Currently, there is significant interest in the exploration and development of IoT systems. Researchers began examining physical systems at an abstract level due to the proliferation of new technologies connected by the Internet [5]. Structured prototyping of systems, subsystems, and components as virtual counterparts is a recent development, although digital twins and various forms of "modeling and simulation" have been around for several decades with the rapid growth of computers. In addition to digital twins, there are numerous terms frequently used in industry, such as—virtual prototype, virtualization, virtual platform, virtual vehicle, aero-DT, virtual replica, and digitalization [6]. While these terms share similarities and are often used interchangeably, they represent distinct concepts and are applied in different contexts depending on the specific requirements of the industry or application. We summarize some common terms with their definitions in Table 3.1.

3.2.2 Digital Twin Market in Vehicular Systems

In 2021, the digital twin industry was valued at \$6.5 billion. By 2030, it is projected to reach \$125.7 billion, growing at a compound annual growth rate (CAGR) of 39.48% [13]. This rapid growth is driven by increasing adoption across various sectors, reflecting the trend towards digital transformation and the use of IoT technologies for improved decision-making and efficiency. The automotive and transportation industries held the largest share of the digital twin market, with aerospace having the second largest share as shown in see Fig. 3.2 [14]. The increasing demand for efficient product design and development has positioned the automotive industry as a leader in prototyping technology among all vehicular sectors. The global digital twins market in the automotive industry is segmented by type, application, technology, and region. In terms of technology, the market is divided into IoT, simulation tools, Artificial Intelligence (AI), Machine Learning (ML), and others, with Fig. 3.3 [15] highlighting the growing importance of simulation tools, i.e., digital twins and other related prototyping technologies. This trend underscores the critical role these technologies play in advancing automotive design and development.

Table 3.1 Prototyping terms definions

Terms	Definition
Driver digital twin (DDT)	A virtual model that mirrors a real driver, incorporating their unique driving data and behavioral patterns [7]. Utilizing real-world data, the DDT system offers both real-time and analytical services in a virtual environment, including driving style analysis and interactive prediction
Virtual vehicle	A digital representation or simulation of a physical vehicle. This concept is widely used in the automotive industry for various purposes, including design, testing, optimization, and performance evaluation. The term "virtual vehicle" was first coined by Sakaguchi et al. [8] for merging control of vehicles on highways. "Virtual platoons" [9] have also been employed for cooperative autonomous vehicles
Virtual platform	In the *cyber* world, virtual prototyping has gained popularity, primarily focusing on digital hardware, embedded software, and System-on-Chips (SoCs). This software-based modeling system, often referred to as a "virtual platform can accurately replicate the operation of specific SoCs or digital hardware. The main objective is to create an abstract model of the hardware as early in the design phase as possible, even before RTL models are available or silicon fabrication begins
Airframe digital twin (ADT)	The US Air Force is exploring a new DT framework to drive innovation in aviation. This physics-based framework provides comprehensive digital data on aircraft performance, safety, and reliability throughout their lifespan, aiding decision-making in acquisition, operation, and maintenance [10, 11]. First introduced at the 2013 ICAF conference, the ADT concept has since been adopted in various aviation prototyping domains, particularly in airframe design [12]

3.2.3 Challenges in Vehicular Prototyping

The evolution of prototyping in vehicular systems has been significantly influenced by the advent of digital twins, virtual platforms, and various prototyping technologies. These innovations have addressed the complex challenges by enabling more accurate simulation and testing environments. Together, these technologies have revolutionized prototyping for vehicular cyber-physical systems (CPS), detecting errors, enhancing efficiency, reducing costs, and accelerating development cycles. Traditional methods struggle to identify these errors before hardware and software assembly. Virtual prototyping offers a solution by allowing early design error detection, facilitating the exploration of CPS configurations, and enhancing understanding of both hardware and software aspects. This approach supports embedded control system engineers in improving overall system functionality.

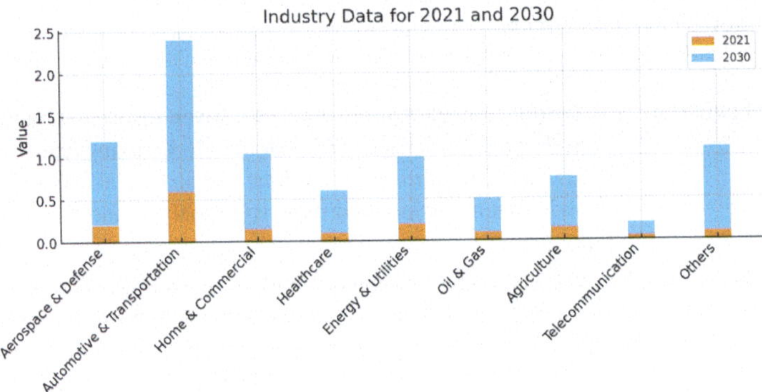

Fig. 3.2 Digital twin market by industry

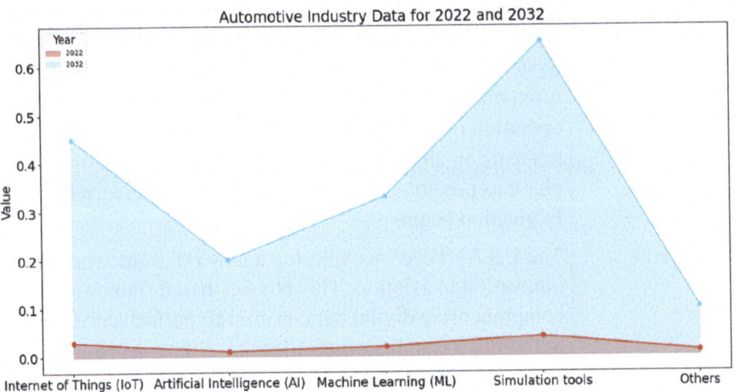

Fig. 3.3 Digital twin automotive market by technology

Current vehicular system-level development (e.g., [16, 17]) primarily targets the physical components or explores domain-specific applications without adequately connecting these aspects to real-time application services. Virtual prototypes address this gap by enabling real-time system monitoring and diagnostics, thereby ensuring the physical systems and application layers work seamlessly together, informed by real-time data analysis and control adjustments [18].

Modern vehicles equipped with advanced technologies like sensors, connectivity, and autonomous driving capabilities, are often cited as prime examples of complex cyber-physical systems. They integrate computational (cyber) processes with physical components, enabling them to interact with their environment and other systems in complex ways. Digital twins and other prototyping technologies are crucial in the vehicular industry, including automotive, aviation, and maritime sectors, for several reasons:

- *Virtual Prototyping*: The creation of virtual prototypes enables engineers to test and refine designs before physical models are built. This accelerates the design process and reduces costs associated with physical prototyping.
- *Simulation and Analysis*: Engineers can simulate various scenarios and analyze the performance of vehicle components under different conditions, leading to more robust and optimized designs.
- *Predictive Maintenance*: Real-time data on vehicle performance supported by advanced prototyping techniques allows for predictive maintenance. This helps in identifying potential issues before they become critical, reducing downtime and maintenance costs [19].
- *Operational Efficiency*: By monitoring and analyzing data for prototypes, operators can optimize routes, fuel consumption, and overall vehicle efficiency.
- *Risk Assessment*: Comprehensive risk assessments are enabled by simulating potential failure modes and their impacts. This helps design safer vehicles and improve regulatory compliance.
- *Regulatory Testing*: Virtual testing can complement physical testing to ensure vehicles meet safety and regulatory standards.
- *Product Lifecycle Management*: Supporting the complete vehicle lifespan, including design, production, operation, and decommissioning, provides a continuous feedback loop for improvements and updates [20].
- *Sustainability*: By optimizing design, operation, and maintenance, these technologies contribute to more sustainable practices, reducing waste and resource consumption.
- *Rapid Innovation*: Prototyping technologies enable rapid iteration and innovation, allowing companies to stay competitive in a fast-evolving market.
- *Customization*: The ability to customize vehicles to meet specific customer requirements improves customer satisfaction and market reach.
- *IoT and Connectivity*: Leveraging IoT technologies to collect and analyze real-time data from connected vehicles enables advanced prototyping, such as simulating vehicle behavior through digital twins, thereby enhancing connectivity and smart functionalities.
- *Autonomous Systems*: In the development of autonomous vehicles, these tools play a crucial role in simulating and testing autonomous systems, ensuring their reliability and safety before deployment.

Overall, digital twins and related technologies are essential in the vehicular industry for driving innovation, improving safety, enhancing operational efficiency, and ensuring sustainable practices.

3.2.4 Model-based Approaches

Model-based approaches are utilized across various advanced fields, including smart assembly, intelligent manufacturing, autonomous driving, aviation training, hydrodynamic testing, marine engine measurement, etc [21]. These methods enhance precision, efficiency, and innovation by enabling simulation, optimization, and real-time monitoring of complex systems. In the case of electric vehicles, the pseudo-2D model [22] is utilized to significantly reduce simulation time and computational effort while maintaining accuracy, enabling efficient cell design, parameter optimization, and battery pack simulation. CARLA, Unity 3D, MATLAB/Simulink, CATIA, ANSYS, etc. are the most commonly used software for 3D and system modeling in the automotive industry. Boeing pioneered the use of computer-aided 3D models (CAD/CAE/CAM) in aerospace design, leading to significant cost savings and reputation gains, with high-fidelity software like CAD, Catia, ANSYS, Siemens NX, and Autodesk driving the industry for over 20 years [23]. Several other studies (e.g., [24, 25]) discussed physics-based models for aero DTs. In marine applications, spectral fatigue analysis, full-ship FEA, hydrodynamic seakeeping, and Monte Carlo models are employed for fatigue damage monitoring and prediction.

Based on the principle of model-based definition (MBD), Liu et al. [26] combined the idea of biological mimicry [27], and proposed a digital twin mimic model (DTMM), which includes multiple DT sub-models (e.g., behavior model, geometry model, and process model). The process forms an in-process model by incorporating measured data into the MBD model, merging it with the designed model, and integrating behavior and context information, resulting in a self-adaptive model. The DTMM represents the product's DT model, encompassing process-oriented changes and integrating process, physical, and geometric information into a 3D model.

3.2.5 Data-driven Approaches

Data-driven approaches play a pivotal role in modern vehicle design, production, and performance optimization. By analyzing large-scale sensor data, telemetry logs, and user feedback, manufacturers can refine engine efficiency, accurately predict maintenance needs, and enhance key safety features. In the automotive industry, a variety of advanced algorithmic solutions are being explored: swarm optimization [28], surrogate models [29], YOLOv4 detection [30], recurrent neural networks (RNNs) [31], and decision trees [32]—all contributing to improvements in autonomous driving, advanced driver assistance systems (ADAS), fault diagnosis, and IoT communications. Meanwhile, the aviation domain has adopted long short-term memory (LSTM) networks [33–35], random forests [36], and convolutional neural networks (CNNs) [37, 38] to streamline data transmission and storage, as well as enable real-time monitoring and predictive analytics. In marine applications, speed loss estimation and fault detection often rely on artificial neural networks (ANN) [39], Q-learning methods [40], and LSTM architectures [41] to ensure reliable operations in challenging maritime conditions. Collectively, these machine learning techniques underscore the growing importance of data-driven strategies for enhancing performance, safety, and cost-effectiveness across multiple transportation sectors.

3.3 Applications in Smart Automotive Domain

The smart automotive system domain has been one of the most extensively explored areas among all vehicular systems for various types of prototyping. Several review papers [42–45] have addressed the application of digital twin technology in the automotive sector from various perspectives. In addressing a major part of **RQ1**, our work examines these applications alongside related prototyping approaches, providing a broader understanding to support emerging research in this field.

3.3.1 Driver Digital Twins

DDT system plays an essential role in intelligently and securely enhancing driver assistance. Its range of applications acts as a vital link between the digital and physical environments, enabling the DDT system to function in a closed-loop mode. Existing applications such as Advanced Driver Assistance Systems (ADAS), shared control, and driver safety monitoring have been under study for decades [45]. The introduction of the DDT system is poised to not only refine the functionality of these existing applications but also to introduce new possibilities that could significantly improve user experiences. Figure 3.4 shows an overview of the DDT concept.

Lio et al. [7] developed DDT to predict personalized lane-change behavior in mixed-traffic environments, enhancing the interaction between automated and human-driven vehicles. The system, evaluated through co-simulation and field tests, demonstrated significant prediction accuracy improvements, detecting lane-change intentions on average 6 seconds in advance with a high level of precision. Ma et al. [46] introduced a novel DDT framework that leverages a transformer-based module and a temporal localization approach to enhance the detection of distracted driving behaviors, integrating driver emotion recognition as a novel element. Their method, tested on three major benchmarks, outperforms existing approaches, confirming its superior accuracy in recognizing driver actions and temporal changes in distracted behaviors.

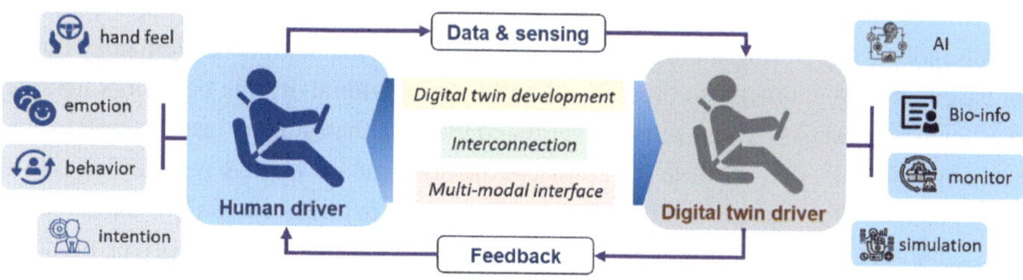

Fig. 3.4 Driver digital twin overview

Apart from the above-mentioned DDT studies, there are various DDT-enabled key applications that do not use the term "driver digital twin" but discuss similar work to enhance the intelligence, personality, and adaptability of driving automation systems at several levels. Several works developed driver models based on behavior and characteristics. Wang et al. [47] proposed a learning-based approach to model driver behavior and reduce speed-tracking errors by classifying drivers into different types using the k-nearest neighbors algorithm and predicting errors with a nonlinear autoregressive neural network. Through simulations and field tests, this approach leads to a 53% reduction in speed error variance and a 3% decrease in energy consumption by compensating for these errors in real-time. Wei et al. [48] calibrated and verified the human-like Wiedemann car-following model using driving simulator data, developing a nonlinear model predictive control (MPC) controller to mimic human car-following behavior. Through testing in three different scenarios, the proposed controller enables autonomous vehicles to exhibit human-like and smooth trajectories, enhancing naturalness in vehicle behavior across various phases and transition zones. Furthermore, a couple of studies [32, 49] explore enhancing Adaptive Cruise Control (ACC) by incorporating personalized driver behaviors and preferences into the system's operation. These approaches utilize machine learning algorithms, demographic data, and behavioral models to tailor ACC features like Stop and Go (S&G), optimizing comfort and efficiency based on individual driving styles and reducing the need for manual interactions with the automated system.

In the area of personalized driving models, Lefèvre et al. [50] proposed a learning-based driver modeling approach that identifies maneuvers and predicts future driver inputs on highways, applying this to provide personalized driving assistance. Utilizing a combination of real data and simulation, the model predicts unintentional lane departures and estimates preferred accelerations, using model predictive controllers to maintain lane integrity and offer personalized adaptive cruise control. Another driver model [51] integrates the Gaussian mixture model and hidden Markov model to predict unintended lane-departure and correction behaviors accurately. This model supports a model-based prediction algorithm and a driver-friendly warning strategy, which, through naturalistic driving data validation, demonstrated a significant reduction in false warnings, decreasing to an average of 3.13% at a 1-second prediction interval. Schnelle et al. proposed a couple of works [52, 53] on driver steering model, integrating a compensatory transfer function with an anticipatory component based on road geometry to predict and replicate individual driving behaviors for safety and efficiency in ADAS. These models, validated using human subject test data and steering wheel angle signals, demonstrate the ability to differentiate between drivers and accurately predict behaviors for various driving maneuvers, including rare collision avoidance scenarios based solely on daily driving data. Li et al. [54] introduced a novel driver model for autonomous vehicles that aligns the cars' movements with the behavioral characteristics of driver owners, based on attributes such as personality, emotion, driving experience, age, and gender. The effectiveness of this model is confirmed through statistical analysis showing a power-law distribution in behavior patterns, and both hardware-in-loop simulations and real driving experiments further validate it. Furthermore, studies [55, 56] have explored driver behavior and driving pattern simulations, validating their findings through case studies using driving simulation experiments.

3.3.2 Virtual Vehicle Applications

Virtual vehicles often form part of DT technology, where a digital replica of a physical vehicle is created and used to monitor and optimize its real-world counterpart. This can include real-time data collection and analysis to predict maintenance needs and enhance performance. Virtual vehicles play an important role in the development of autonomous vehicles. They are used to simulate various driving scenarios and test the vehicle's autonomous systems, ensuring that they can handle a wide range of situations safely and effectively.

Debada et al. [57] proposed an innovative distributed coordinating system for connected autonomous vehicles (CAVs) based on virtual vehicles, enabling them to position virtual vehicles to share intended maneuvers and request cooperation. Simulation results demonstrate that this framework significantly enhances performance, especially in a three-legged single-lane roundabout, although mixing CAVs with unconnected vehicles leads to performance degradation and highlights the need for effective coordination strategies. Leitao et al. [58] developed a framework for the control and simulation of numerous autonomous agents in real-time interactive systems, specifically focusing on autonomous vehicles and pedestrians within the DriS simulator. Their approach, which employs a scripting language based on Grafcet, allows for the imposition of both short-term orders and long-term goals, enabling the model to handle both environmental traffic and controlled vehicles effectively. Egerstedt et al. [59] proposed two model-independent solutions for controlling wheel-based mobile platforms using a virtual vehicle approach, where the reference point's motion on the desired trajectory is determined by a differential equation with error feedback. These solutions, being robust to errors and disturbances, employ stable control algorithms that function as proportional regulators with arbitrary positive gains, as evidenced by experimental results.

Another similar study [60] explored the coordination control of leader-follower systems by implementing a virtual vehicle strategy, which minimizes the amount of information needed from the leader. This method stabilizes the virtual vehicle to the leader, providing a reference velocity for the follower's synchronization control, demonstrating uniformly semi-globally practically asymptotically stable closed-loop errors with only position/heading measurements. An existing autonomous vehicle (AV) data set was used in a tutorial article [61] to analyze a suggested framework for building, parameterizing, and verifying a virtual vehicle environment. It reviews several open-source and commercial simulation tools, discusses the selection of AV data sets for validation, demonstrates recreating real-world scenes in simulation, and suggests metrics for analyzing AV-perception algorithms using data from both virtual and real-world environments.

Popular navigation systems for the Internet of Vehicles (IoV) assist drivers in planning routes and navigating real-time road congestion. However, these systems may inadvertently cause congestion on alternate routes, so Lei et al. [62] introduced a virtual vehicle concept in IoV to study non-atomic routing and uses strategic concession games to minimize both individual and total travel times, showing that cooperation significantly reduces efficiency loss compared to non-cooperative approaches. Liu and Ma [63] developed a method for collecting and archiving artery data in real-time at the University of Minnesota, which gathers high-resolution event-based traffic data from multiple intersections to estimate time-dependent travel times using a virtual probe vehicle algorithm. Field studies demonstrated that under different traffic scenarios, the suggested approach can produce precise time-dependent travel times by self-correcting errors through the synchronization of virtual and real probe vehicle trajectories. Tsanakas et al. [64] proposed a novel approach for generating Virtual Vehicle Trajectories (VVT) to improve emission modeling accuracy, addressing the limitations of traditional VVT methods that simplify vehicle kinematics. Empirical evaluations show that their method enhances emission estimations when compared to traditional approaches, especially under specific experimental conditions. Moreover, the application of virtual vehicles has been instrumental in several studies including the role of virtual vehicle cab [65], cooperative maneuver planning [66], predictive approach for automated vehicle control [67], and connecting transport [68].

3.3.3 Virtual Platforms for Automotives

This virtual platform serves as a customizable environment for design optimization and software development. This approach has proven advantageous for early software and hardware development, verification, validation, and hardware-software co-design [69, 70]. Moreover, virtual platforms have been supported by various commercial frameworks [71]. The objective of this approach is to create a simplified model of the hardware device that provides a configurable platform at an early stage in the design process. This is advantageous even before silicon prototypes or detailed register-transfer level (RTL) models are developed. Such platforms are invaluable for software development and design optimization. Additionally, it has proven effective in the early stages of software and hardware development, verification, and validation, and in facilitating hardware-software co-design [69, 70]. Pro-SiVIC is a virtual platform simulation software by ESI Group designed to simulate vehicle sensor systems. It is particularly useful for testing autonomous vehicles, and ADAS [72, 73], [74], allowing for the creation of realistic 3D simulations of various scenarios, including road users and environmental interactions. The platform helps in sensor development by simulating how sensors perceive different environments, which assists in integration and testing processes, ultimately enhancing the development of safety systems in vehicles. It runs on Windows 10 and features tools for trajectory design, Python API integration, and 3D model imports. The R-Car Virtual Platform (VPF) [75] is a noteworthy simulation

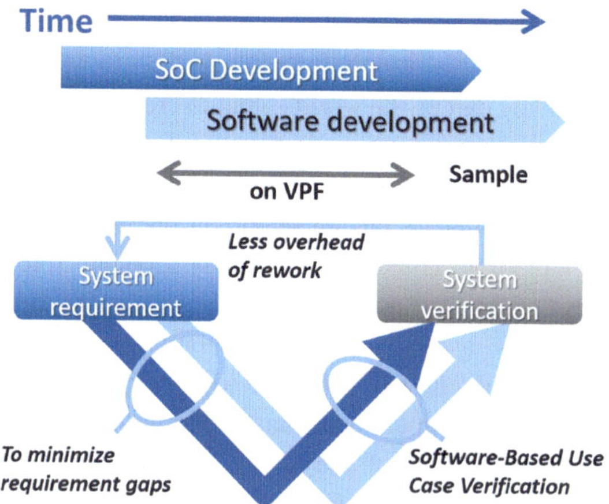

Fig. 3.5 R-Car Virtual Platform [75] overview

environment that enables software design and system verification without a physical device, emulating the R-Car device's functions at the register interface level. It supports a seamless transition from pre-silicon to post-silicon development, providing binary compatibility with the actual device and facilitating early system verification, regression testing, and cooperative operation of device and control software. Figure 3.5 shows their platform overview as it provides a generic outlook of the virtual platform concept for automotives.

Strobl et al. [76] discussed the growing adoption of virtualization in embedded devices like avionics systems and mobile phones, highlighting its first significant use in video game consoles. Despite its potential to consolidate numerous Electronic Control Units (ECUs) into a few Domain Controller Units (DCUs), virtualization has yet to be employed in automotive electronics, a gap this paper aims to address by presenting the benefits of automotive virtualization, particularly for DCUs. Safari et al. [77] proposed an integrated framework for Virtual Verification and Validation (VVV) that simulates and emulates a complete automotive system at three levels: System on Chip (SoC), ECU, and system level. This framework enhances the automotive V-cycle by supporting simultaneous software and hardware development, provides detailed methods for debugging AUTOSAR applications, and offers tools for system tracing, profiling, and debugging, ultimately reducing the development cycle as system complexity increases. A Virtualized Automotive Display System, as proposed by Lee et al. [78], introduces a system called VADI to virtualize a GPU and its attached display device for modern vehicles. This system manages two execution domains for automotive control and in-vehicle infotainment (IVI) software, ensuring isolated GPU operation and compliance with ISO safety standards by maintaining a minimum of 30 fps in failure scenarios and achieving 60 fps under synthetic workloads.

3.3.4 Autonomous Driving Test

Autonomous driving is a key part of the world's goal to create smarter, safer transportation systems. There have been increased efforts in the recent past contributing to autonomous driving [79–81]. However, Autonomous Vehicles (AVs) will be subject to inherent risks that cannot be completely mitigated, raising important questions about risk management, including which risks can be accepted [82]. DTs play a crucial role in the development, testing, and deployment of AVs by providing a virtual representation of physical assets, processes, and systems, enabling comprehensive simulation and analysis. The need for DTs in autonomous driving technologies stems from two major concerns: safety and cost. Tests conducted in a simulation environment and transferring the learned driving knowledge to the real world form a critical aspect of autonomous driving DTs. In the context of AVs, DTs can represent the vehicle, its environment, or the broader transportation ecosystem, continuously updating with real-time data for accurate simulation and optimization [83]. The vehicle model digitally replicates mechanical, electrical, and software systems, while the environment model includes other vehicles, pedestrians, and objects detected by GPS, LIDAR, RADAR, and other sensors. Key challenges include real-time data integration, maintaining high-fidelity models, ensuring cybersecurity, and scaling DT infrastructure to accommodate evolving AV systems and environments.

In the autonomous vehicle space, DTs enable testing and refinement before physical prototypes are built. This reduces the cost and accelerates development. A survey by Hu et al. [84] explored how simulation technologies like sim2real, DTs, and parallel intelligence (PI) contribute to autonomous driving. It reviews the latest algorithms, models, and simulators, detailing the progression from sim2real to DTs and PI. Additionally, the article discusses the challenges and future prospects for these technologies in enhancing autonomous driving. Niaz et al. [44] proposes a framework using a DT to test connected autonomous driving in controlled environments, utilizing DT maneuvers to simulate real driving tests in virtual complex road scenarios. It examines DTs from various perspectives, including their foundations, development, industrial applications, cyber-physical integration, and core components, highlighting both their applications and the challenges they face. Guo et al. [85] reviews the integration of DTs into the design and deployment of Internet of Vehicles (IoV) systems, identifying it as an understudied area. It explains the benefits DTs can offer to IoV system designers and implementers and outlines various challenges and opportunities for future research in this field. This study [86] provides a comprehensive review of DT technology in smart electric vehicles, bridging individual research efforts and offering a technically informed and neutral overview. It systematically explores the inception and evolution of DT technology, highlighting contributions to smart vehicle systems and addressing challenges. The review covers domains such as advanced driver assistance, battery management, power electronics, power drive systems, autonomous navigation, and vehicle health monitoring. It also examines the techno-socio-economic impact and potential obstacles to further development, presenting an extensive and holistic perspective on DT applications in smart electric vehicles.

Almeaibed et al. [87] addressed the challenges and solutions for ensuring safety and security in autonomous vehicle systems using DT. It aims to establish a standard framework for vehicular DTs to enhance data collection, processing, and analytics. To illustrate the effectiveness of this approach, the article includes a case study of a vehicle follower model where radar sensor data is manipulated to simulate a collision scenario. Using the Unity game engine, Wang et al. [83] proposed a simulation architecture for automated and connected vehicles. The simulation comprises a physical and a digital environment, with the latter split into three layers: external tools such as Python, MATLAB, SUMO, and AWS to improve functionality, Unity scripting for software, and Unity game objects to simulate hardware. A case study on personalized adaptive cruise control (P-ACC) demonstrates the effectiveness of this approach, showing how the ACC system can be tailored to individual driver preferences using cloud computing. Thonhofer et al. [88] developed a DT-based information model to provide high-quality, machine-usable data for automated driving functions, addressing the challenge of formalizing complex and incomplete information. By proposing and validating a system architecture through three implementations, they demonstrate its effectiveness in supporting decision-making at a highway tunnel construction site in Austria and across the Test Bed Lower Saxony in Germany.

3.3.5 Exploration Platforms and Simulators

Automotive security has experienced considerable growth and still possesses ample potential for further enhancements. This progress became particularly noticeable when various exploration platforms and simulators significantly impacted and contributed to the field. Exploration platforms and software-based simulators enable operators to replicate or simulate phenomena anticipated to occur in actual performance under test conditions, allowing seamless exploration of the intended target components and systems These platforms vary widely in their approach and technology. Some align more closely with the concepts of DTs or virtual reality frameworks. Others rely on a model-based approach, which enables operators to replicate or simulate phenomena anticipated to occur in actual performance under test conditions, allowing seamless exploration of the intended target components and systems [89]. They utilized simulations to predict and analyze the behavior of automotive security systems under different conditions and threats. Furthermore, exploration platforms help to seamlessly explore vehicular electronics, automotive system-level scenarios, traffic simulations, etc. Table 3.2 gives an overview of several exploration platforms and automotive simulators focusing on various aspects of the automotive and transportation domain. The platforms discussed here are those that have either made a significant impact in the research domain or are part of emerging research. Commercial offerings, i.e., established modeling software packages, are not included in this analysis.

Table 3.2 List of exploration platforms

Title	Application
Ahura [90]	An open racing car simulator controller based on heuristics, featuring five modules: the dynamic adjuster, opponent manager, steer controller, speed controller, and stuck handler
AutoDRIVE [91]	A simulator for scaled autonomous vehicle research and education
Carla [92]	A comprehensive framework for the development, training, and validation of autonomous driving systems
IVE [93]	An immersive virtual environment to study safety vulnerabilities in automotive systems in response to cyberattack concerns
Metropolis [94]	Exploring the design space of automotive platforms with a focus on software distribution, architecture, and network setup
Sumo [95]	A portable, open-source, microscopic, and continuous multi-modal traffic modeling program that can manage large networks
TORCS [96]	A modern, modular, highly portable multi-player, multi-agent racing car simulator
VANETsim [97]	An open-source simulator for analyzing security and privacy concepts in Vehicular Ad Hoc Networks
ViSE [98, 99]	A digital twin exploration platform focusing on automotive functional safety and various security vulnerabilities by exploring use cases
ViVE [100, 101]	Modeling and simulation of vehicular electronics through various system-level use cases
VISSIM [102]	Modeling, simulating, and analyzing traffic and transportation systems

3.4 Applications in Smart Aviation Domain

In the world of IoT-based smart aviation, the concept of a digital twin represents a groundbreaking shift. DTs create a digital replica of physical aircraft, incorporating real-time data to model, predict, and enhance the performance and maintenance of the aircraft throughout its lifecycle. By interfacing with IoT, DTs facilitate enhanced decision-making, improve operational efficiency, and contribute to the safety and sustainability of aviation operations. Several studies [103, 104] have conducted surveys on the application of DTs throughout the lifecycle of aviation systems. A simplified overview of ADT is shown in Fig. 3.6. We explore the principal applications of prototyping, focusing primarily on aircraft design, health monitoring, and the UAF field.

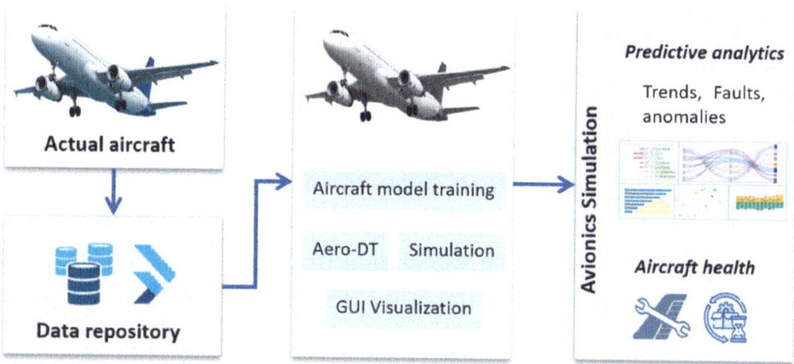

Fig. 3.6 ADT model simplified overview

3.4.1 Aircraft Design and Assembly

The design and assembly of aircraft encompass intricate processes marked by significant complexity and evolving methodologies. Initially grounded in static strength principles, aircraft structural design has progressively incorporated concepts such as safe life, damage tolerance, and stand-alone tracking [105–107]. Despite these advancements, challenges persist, including the high degree of variable coupling, insufficient data, and difficulties in obtaining performance indicators. Given that aircraft are highly complex products, their assembly processes are characterized by high complexity, a dynamic nature, uncertainty, and frequent maintenance requirements.

Managing the lifecycle of complex, safety-critical systems from design to operation remains challenging due to the need to track dependencies, changes, and updates. Therefore, Bachelor et al. [108] discussed how model-based DT and thread concepts can improve the model-based design process by addressing issues related to IT infrastructure integration and tool interoperability. A case study on the design of an ice protection system for a regional aircraft is used to illustrate these concepts. AeroVR [109], an immersive aerospace design environment, facilitates the aerodynamic design process through interactive visualization of geometry and performance. The paper breaks down the design into functional structures, defines primary and secondary tasks, explains the system implementation, and uses an engine compressor blade design study as a prototype to validate the interface's expressiveness and usability.

In the line of the aircraft assembly process, Demartini et al. [110] illustrated the critical role of systems that combine manufacturing operation management (MOM) and product lifecycle management (PLM) in achieving Digital Manufacturing by enabling seamless information exchange through DTs and Digital Threads. Developed within the Siemens Industry Software framework under the AirGreen 2 project, this approach optimizes production operations, scheduling, and performance analysis, thereby enhancing cost efficiency, time management, and quality in manufacturing processes. A couple of studies by

Cai et al. [111, 112] emphasized that an aircraft's final assembly is crucial for determining an aircraft's reliability and service life. They highlight the challenges in quality management, such as handling multi-source heterogeneous data and non-uniform data exchange formats, and propose using the aircraft DT to enhance quality management by creating a unified data source and employing data mining tools to accurately identify quality issues. A different approach was taken by Guo et al. [113] to improve assembly coordination technology by developing a working mode based on the Digital Coordination Model (DCM) instead of hard master tooling. They demonstrated that this DCM-based mode significantly enhances assembly accuracy, coordination accuracy, and efficiency, increasing assembly efficiency by about 40%.

3.4.2 Aircraft Health and Maintenance

Aircraft health is critical not only for ensuring safety in aviation but also for optimizing operational efficiency and minimizing maintenance costs. As aviation technology evolves, maintaining the robustness and reliability of aircraft systems has become more complex and demanding. In this context, the advent of the DTs marks a significant leap forward. Through continuous updates and simulation capabilities, DTs provide a powerful tool for managing the lifecycle of aircraft, making aviation safer, more efficient, and economically viable. Integrated Vehicle Health Management (IVHM) facilitates Condition-Based Maintenance (CBM) by monitoring, diagnosing, and predicting the health and performance of aerospace systems [114]. In aerospace, DTs enhance IVHM by predicting issues before they occur, optimizing maintenance schedules, and reducing operational costs. The functioning of an IVHM system is guided by the OSA-CBM framework, which outlines steps for data collection, processing, and analysis to monitor, diagnose, and predict the health of aircraft components as shown in Fig. 3.7 (adapted from [114, 115]). The DT enhances this process by providing a virtual representation of the aircraft, ensuring the reliability of condition monitoring, diagnosis, and prognosis results. These insights are then used to schedule maintenance programs efficiently.

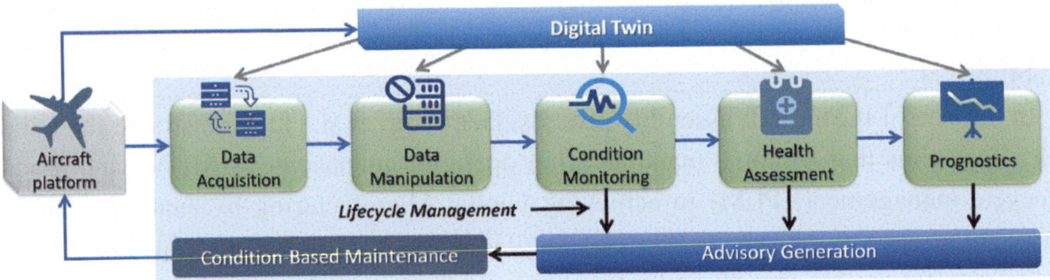

Fig. 3.7 IVHM framework with digital twin

Various approaches and components of health monitoring utilize DTs to comprehensively evaluate aircraft health. Li et al. [116] proposed using a dynamic Bayesian network to create a versatile probabilistic model for diagnosis and prognosis, advancing the DT vision in aircraft health monitoring. By integrating physics models and handling various uncertainties, this approach effectively predicts and tracks crack growth, reducing time costs through a modified network structure and particle filter implementation. Zakrajsek et al. [117] developed a DT model to predict aircraft tire wear at touchdown, incorporating a physics-based tire wear equation and using Monte Carlo and Cotter sensitivity analyses to assess uncertainty. The DT model calculates the Probability of Failure (POF) under various conditions, showing potential benefits for cost savings and tire health monitoring, with future enhancements recommended for more accurate life predictions. A noteworthy work [118] on IVHM processes described a data-driven approach to synthesizing and analyzing DTs of avionics functions using machine learning classification algorithms to address issues such as function loss and behavioral indications, ultimately enhancing the design and development stages of the avionics lifecycle.

In the field of Structural Health Management (SHM), the precise identification and assessment of aircraft damage under diverse flight conditions are enabled by advancements in interdisciplinary integrated tools. SHM systems provide real-time estimates of the amount of damage and safe load-bearing capability by utilizing multiphysics models, sensor data, and inputs from in-service vehicles. A study [119] employed guided wave responses and a genetic algorithm optimization procedure to estimate damage size, position, and orientation by comparing signal responses between damaged and undamaged structures. Similarly, Lai [120] presented a DT-based structural health monitoring framework combining measurement and computational data. The framework involves four steps: using AI-driven load identification, applying multi-fidelity surrogate models, developing an online rain flow counting algorithm, and fusing data into a 3D scene. The MCC-DT framework is demonstrated using an aircraft model to verify its applicability and effectiveness. Instead of focusing on individual aircraft, some studies [121–123] proposed frameworks for fleet-wide monitoring, diagnostics, and health management, using a multi-level approach that begins with threshold exceedance monitoring and progresses to fault detection for isolated problematic engines. Various techniques for the identification, isolation, and quantification of faults are compared and tested on fleet data, and the limitations of both physics-based and machine-learning techniques are discussed to highlight the need for effective fleet diagnostics.

For intelligent predictive maintenance, numerous researchers have achieved significant advancements through the use of DT prototyping, making valuable contributions to the field. Liu et al. [124] reviewed the framework for developing a DT integrated with industrial IoT to enhance the autonomy of aerospace platforms, highlighting the importance of data fusion techniques. They also discussed the Information flow from unprocessed data to advanced decision-making, emphasizing the roles of fusion between sensors, models, and models with sensors in aircraft predictive maintenance. Xiong et al. [33] studied an aero-engine predictive maintenance framework driven by DT, focusing on mining the implicit digital twin (IDT) model and verifying its validity through consistency evaluation of virtual and real data assets. They demonstrate the effectiveness of combining a data-driven approach with an LSTM deep learning model, achieving high prediction accuracy with an RMSE of 13.12 when 80% of the dataset is used for training, surpassing other experimental schemes. Furthermore, various technologies can enhance remote collaboration and increase the efficiency of aircraft maintenance, such as augmented reality (AR) [125], DT by Rolls Royce [126], and component-level simulations [127].

3.4.3 UAV Field

With the growing size, participation, and volume of traffic, the unmanned aerial vehicle (UAV) market has emerged as one of the aviation industry's most dynamic sectors. At the moment, some emerging research is being conducted on UAV DTs, which are primarily used as auxiliary tools in other DT domains [104]. A noteworthy aspect of these research works is that they have been conducted by the same group of researchers across multiple papers. Lv et al. have a couple of works where (1) they explored the application effects and limitations of UAVs in 5G/B5G communication, demonstrating that their proposed algorithm significantly enhances estimation accuracy and robustness compared to traditional methods [128], and (2) demonstrated that DTs in UAVs can significantly improve the speed, accuracy, and efficiency of delivering medical resources during COVID-19 prevention and control, with their model achieving 95.58% accuracy [129]. By employing the use of deep reinforcement learning (DRL), a group of researchers published three works, all of which used deep reinforcement learning (DRL). The first one [130] proposed a novel DT-based intelligent cooperation framework for UAV swarms, integrating a high-fidelity DT model and a machine-learning algorithm to optimize and control UAV behavior. The second one [131] proposed a DT-enabled DRL training framework, specifically an actor-critic DRL algorithm, to improve the practicality and performance of flocking motion in multi-UAV systems. The third one [132] proposed a new multi-agent DRL i.e., a MADRL method called DNQMIX, along with a DT-driven training framework, to enhance the efficiency and performance of multi-UAV cooperative target search in dynamic environments. Moreover, Li et al. [133] and Tang et al. [134] also utilized the DRL approach to improve task assignment and path planning efficiency in multi-UAV systems.

Another group of researchers had a couple of works where (1) they have introduced a novel DT simulation platform for multi-rotor UAVs that integrates Unity, ROS, Matlab, and SimulIDE to verify and track the UAV's lifecycle, providing multi-dimensional and multi-scale simulations to improve development efficiency and functionality [135] and (2) presented the first integration of DT, 5G, cloud platform, and VR technologies for UAV autonomy development, proposing "DTUAV"—a comprehensive DT framework and demonstrating its effectiveness through multiple experiments in real-time monitoring, VR interaction, and remote supervision [136]. For the Internet of vehicle networks, Hazarika et al. [137] proposed a DT-assisted intelligent delay-sensitive task offloading scheme utilizing UAVs as relays to improve resource allocation for delay-intolerant tasks. They introduced a multi-network DRL-based resource allocation algorithm (RADiT), which significantly enhances energy efficiency and reduces network delay compared to the soft actor-critic (SAC) algorithm and a non-DRL greedy approach. Moreover, there are significant other UAV-focused works e.g., vertical take-off and landing (VTOL) [138], path planning [139], AI-based framework [140], and warehouse management system [141].

3.4.4 Avionics Simulation and Security Issues

Simulation of various avionics functions and the analysis of cyber-attacks on avionics networks have been crucial areas for digital twin exploration. Kandaz et al. [142] developed a simulation-based DT of a high-speed turbomachine, specifically a fan with a nominal speed of 16,500 rpm, for the aerospace industry, utilizing multi-physics engineering simulations and validated via experimental data. This DT, capable of performing various operational scenarios and stress tests, demonstrates significant potential in enhancing operation and maintenance practices through dynamic reduced-order models and what-if analyses, benefiting O&M teams, asset owners, and OEMs. Fraser et al. [143] investigated the vulnerability of UAVs to contemporary cyber threats and proposed remedies utilizing data-driven techniques and DT structures to detect anomalies and intrusions in real-time. These concepts were validated by applying novelty detection to GPS spoofing attacks on UAV flight data. Various machine learning models—both deep learning and classical—were employed to determine which models were best suited for the proposed design, yielding promising results. Another work by Kuleshov et al. [144] presented a novel academia-industry collaboration between Purdue University and Boeing, aimed at addressing vulnerabilities and cyberattacks on airplane avionics networks through a DT model. This collaboration, part of the Data Mine Corporate Partners program, allowed students to simulate and analyze potential cyber threats on general-purpose computers, providing valuable insights and fostering the development of new cybersecurity approaches in aviation.

3.5 Applications in Maritime Vehicle Domain

Maritime transport is the most economical means of transporting people and cargo, serving as the backbone of international trade. Much of what we possess has traveled by sea, even though maritime transportation often does not receive adequate recognition. Numerous innovations in this field have had a profound impact on human life. From being wind-powered to now utilizing batteries, maritime transport has undergone significant advancements. The use of DTs is one such innovation. Although most DTs in the maritime industry have been primarily studied for ships [145], we will highlight notable works here that explore applications across various types of maritime vehicles.

3.5.1 Ship Design

Testing full-scale ships is prohibitively expensive, so the principle of similitude is employed, using scaled models to test designs. While cost-effective and easier, this method lacked accuracy. With the advent of Computer Aided Engineering (CAE), engineers began using Finite Element Analysis (FEA) software to simulate and validate structural and fluid dynamics. DTs enable virtual testing and validation of new designs and modifications, reducing the need for physical prototypes and accelerating innovation and deployment. DTs also allow engineers to simulate various environmental conditions and operational scenarios to optimize performance and make necessary adjustments. Figure 3.8 shows DT activity in maritime vehicles, showing the ship as an illustrative example.

Fonseca et al. [146] proposed creating an open standard for DT data based on lessons from existing literature and previous attempts at standardizing ship data. The proposal aims to address challenges and guide the future development of services, networks, and software for DTs, with a specific application to a research vessel. Mombiela et al. [147] proposed an algorithm for designing ship power systems during the preliminary design phase. It includes an embedded control for Energy Management Systems (EMS) to optimize cost, availability, safety, and emissions. The study evaluates different power system alternatives, such as full electric propulsion and fuel-based energy producers, using various performance and improvement indicators. A case study on an Offshore Supply Vessel compares battery and GENSET power contributions, analyzing battery performance and control metrics. The authors claim that the algorithm serves as a foundation for future hybrid power plant designs. Menges et al. [148] proposed enhancing autonomous surface vessels (ASVs) with a DT framework to improve situational awareness and decision-making. They integrate real-

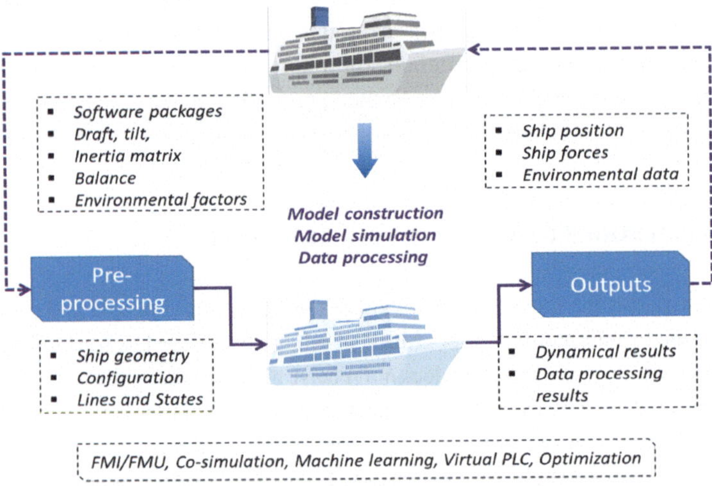

Fig. 3.8 DT-driven approach for prototyping in ships

time AIS and synthetic Lidar data for sophisticated target tracking and apply a predictive safety filter based on nonlinear model predictive control (NMPC). This DT, built using the Unity game engine, enables predictive, prescriptive, and autonomous capabilities for safer and more efficient maritime operations. Some other works include ship architecture for optimization [149], open simulation platform [150], and cruise ship DT [151].

3.5.2 Navigation and Operations

Hasan et al. [152] introduced a method to enhance the reliability and efficiency of autonomous surface vessels (ASVs) using predictive DT and state estimation techniques. A predictive algorithm is developed from measured data to create a discrete-time state-space model, forecasting the system's future response. Tests conducted on the ASV by Maritime Robotics show the DT's potential as a decision support tool for ASV maintenance. Liu et al. [153] proposed an approach for improving the safety and intelligence of the Maritime Transportation System (MTS) using DTs and the Internet of Things (IoT). This work analyzes historical data from the Maritime Silk Road and constructs a DTs model with relay nodes to enhance data transmission and security. Simulation experiments validate the model's security performance, showing that interference helps maintain information confidentiality, and increasing relays improves system performance and reduces outage probability. This study concludes that the proposed model offers excellent transmission and security, supporting future intelligent and secure maritime transportation. Kaklis et al. [154] developed a DT to improve the operational efficiency of the vessel from the context of AI and Industry 4.0. Their work focuses on estimating Fuel-Oil-Consumption (FOC) for ships using a novel method with a reduced feature set to predict the main engine's rotational speed (RPM). It integrates B-Splines with deep learning architectures and compares them with existing regression techniques. The goal is to improve efficiency and environmental compliance in ship operations by using these models within a cutting-edge information system. The paper shows the effectiveness of their estimators by comparing them with actual FOC data. Additionally, a few more emerging research [149, 155, 156] focus on autonomous navigation systems.

3.5.3 Fault Diagnosis in Autonomous Vessels

Bhagavathi et al. [157] introduced a DT-driven fault diagnosis method for autonomous surface vehicles, utilizing a graphical model and an adaptive extended Kalman filter algorithm. The authors claim that, unlike traditional methods, this new algorithm directly estimates fault magnitudes from sensor data. This approach is also tested on Otter by Maritime Robotics and the DT receives real-time data, estimates actuator faults, and visualizes results via a web application. Simulation and experimental results demonstrate accurate fault detection and estimation. Hasan et al. [158] introduced a DT-based method for diagnosing faults in autonomous ships. The system collects data from ship sensors, analyzes it, calculates fault parameters, and displays them in real time. Using a dynamical model and an adaptive extended Kalman filter (AEKF) algorithm, it estimates fault severity. Yet again, this method is tested on Otter by Maritime Robotics. The method shows potential in improving ship condition monitoring, as demonstrated through simulations and practical implementation.

3.5.4 Underwater Vehicles

Similar to UAVs in aviation, DT technology has also been applied to underwater vehicles in recent publications. With the use of big data and machine learning approaches, computational fluid dynamics, and fluid operating conditions, a DT escort system model has been proposed by Gao et al. [159] to enhance the performance of underwater unmanned vehicles (UUVs). This research provides innovative approaches to maneuver early warning, wake tracking, UUV intelligent obstacle avoidance, and flow field detection, addressing issues of disconnection and loss in complex underwater environments. Martynova et al. [160] are developing a DT of an autonomous underwater vehicle to assess its effectiveness in searching for bottom objects by analyzing factors such as area size, inspection swath width, detection principles, bottom topography, and marine environment state. Using the Monte Carlo method, the simulation reproduces the movement of detection means, view sectors, and the sequential selection of detection methods based on environmental conditions and object classification characteristics.

Furthermore, several works have been done on prototyping for autonomous gliders. Hu et al. [161] proposed a novel underwater simulation platform based on DT technology, using a blended-wing-body underwater glider (BWBUG) to explore high-fidelity digital modeling in the marine environment. This platform, incorporating sensor data with a PID control algorithm and robot operating system (ROS), achieved enhanced pitch control, reducing overshoot to 0.06% and settling time to 29.5 seconds, demonstrating the potential for safety evaluation, design enhancement, and performance prediction in the design of the next generation of marine vehicles. Wen et al. [162] proposed an improved artificial potential field method for multi-agent Autonomous Underwater Glider (AUG) systems, leveraging the Maritime IoT and edge computing to enable collaborative control by a leader AUG. This approach includes a behavior-based path optimization to overcome local optima and utilizes a DT model for real-time monitoring, enhancing coordination and communication among multi-AUG groups in challenging underwater environments. Another work by Yang et al. [163] introduced a DT-driven architecture to address the challenges of Full Lifecycle Management (FLM) and Rapid Individualized Design (RID) for underwater gliders, tackling interdisciplinary complexities and environmental uncertainties. This architecture includes virtual verification, design optimization, digital modeling, and real-world implementation, demonstrated through a case study with the Petrel UG developed in China, showcasing its feasibility for other autonomous underwater vehicles as well.

3.6 Future Research Directions

As prototyping technology continues to evolve, its potential applications in the vehicular sector expand significantly. To harness the full capabilities of various prototyping technologies, future research should address several key areas. In addressing **RQ3**, this section explores how insights from existing approaches can guide future research directions, fostering more effective and sustainable development of next-generation vehicular infrastructure.

3.6.1 Integration with Emerging Technologies

Artificial Intelligence and Machine Learning

The integration of AI and ML with DT technology has the potential to significantly enhance predictive maintenance, anomaly detection, and system-level optimization in vehicular systems. By learning from historical and real-time data streams, AI-driven models can improve fault diagnosis, performance forecasting, and adaptive control strategies. Future research should focus on developing scalable learning algorithms that can efficiently process the high-volume, high-velocity data generated by DTs, while also addressing challenges related to model interpretability, robustness, and real-time deployment in safety-critical environments.

5G and Beyond

The deployment of 5G wireless networks introduces ultra-low latency, high bandwidth, and massive device connectivity, which are essential for real-time DT operation. These capabilities enable frequent synchronization between physical vehicles and their digital counterparts, supporting fine-grained state updates and rapid feedback. Future research should investigate how 5G and emerging beyond-5G (B5G/6G) technologies can support distributed DT architectures, particularly in highly dynamic vehicular environments involving cooperative driving, edge intelligence, and V2X communication.

Blockchain for Security and Data Integrity

Ensuring the security, integrity, and traceability of data exchanged within DT-based prototyping systems is critical, especially in distributed and collaborative environments. Blockchain technology offers decentralized mechanisms for secure data sharing, immutable logging, and trust management across multiple stakeholders. Future studies should explore lightweight and scalable blockchain frameworks that can be integrated with DTs to protect data transmission, ensure provenance, and support access control, while minimizing computational and latency overheads.

3.6.2 Enhanced Modeling and Simulation Capabilities

Multi-Scale and Multi-Physics Simulation

The growing complexity of modern vehicular systems necessitates advanced modeling approaches capable of capturing interactions across multiple spatial and temporal scales, as well as diverse physical domains such as mechanical, electrical, thermal, and control systems. Future research should focus on developing multi-scale and multi-physics simulation frameworks that allow seamless integration of component-level models with system-level behavior. Such approaches can improve the fidelity and predictive power of DTs, enabling more accurate performance evaluation and design-space exploration.

Human Factors and Behavioral Modeling

Incorporating human factors into DTs is essential for developing intelligent, user-centric vehicular systems. Driver behavior, decision-making patterns, and interaction with automated systems significantly influence overall system performance and safety. Future research should aim to develop accurate driver digital twins that model individual behaviors, preferences, and cognitive states. These personalized models can support adaptive human–machine interfaces, improved driver assistance systems, and safer transitions between manual and automated driving modes.

3.6.3 Interoperability and Standardization

Cross-Domain Integration

Vehicular systems increasingly operate within interconnected transportation ecosystems that span multiple domains, including automotive, aviation, and maritime sectors. Research should investigate interoperable DT architectures that enable seamless integration across these domains, supporting unified modeling, simulation, and control. Such cross-domain DTs can facilitate coordinated mobility services, multimodal transportation planning, and holistic safety analysis across heterogeneous platforms.

Standardization Efforts

The large-scale adoption of DT technology depends on the availability of well-defined standards for data representation, communication protocols, and modeling frameworks. Ongoing and future research should contribute to standardization efforts by aligning DT implementations with emerging industry and international standards. Establishing common interfaces and reference architectures will enhance interoperability, reduce integration costs, and accelerate technology transfer from research to practice.

3.6.4 Sustainability and Environmental Impact

Green Technologies and Energy Efficiency

DT-enabled prototyping offers significant opportunities to improve the environmental sustainability of vehicular systems. Virtual models can be used to optimize energy consumption, reduce emissions, and evaluate the integration of alternative powertrains and renewable energy sources. Future research should explore how DTs can support eco-design, energy-aware operation, and real-time monitoring of environmental performance throughout a vehicle's operational lifecycle.

Lifecycle Analysis and Circular Economy

Beyond operational optimization, DTs can facilitate comprehensive lifecycle analysis, encompassing design, manufacturing, operation, maintenance, and end-of-life stages. Future studies should investigate DT-driven approaches for supporting circular economy practices, such as predictive maintenance for extended asset life, remanufacturing planning, and efficient recycling. By providing data-driven insights across the lifecycle, DTs can help reduce waste and promote sustainable industrial practices.

3.6.5 User-Centric Applications and Societal Impact

Personalized Mobility Solutions

DT technology enables the development of personalized mobility services tailored to individual user needs, preferences, and contexts. Future research should examine how DTs can support adaptive transportation systems that respond dynamically to user behavior, traffic conditions, and environmental constraints. Such systems can improve travel efficiency, user satisfaction, and accessibility in both urban and rural settings.

Ethical and Societal Implications

As DT-based prototyping becomes increasingly integrated into transportation and mobility infrastructures, it is important to address associated ethical and societal concerns. Future research should examine issues related to data privacy, algorithmic transparency, and equity of access, as well as the broader impact of DT adoption on workforce roles and societal structures [164]. Addressing these challenges is essential for ensuring responsible and inclusive deployment of DT technologies. By advancing these research directions, the transformative potential of DT technology in the vehicular domain can be fully realized, enabling smarter, safer, and more sustainable transportation systems.

3.7 Conclusion

Digital Twins and other related prototyping methods have all had a major impact on the development of prototyping in CPS. One of the most common examples of a CPS is a modern vehicle that has been outfitted with sophisticated sensors, connections, and autonomous driving features. They combine physical elements with computing (cyber) processes, allowing for intricate interactions with their surroundings and other systems. We have explored various vehicular prototyping concepts and provided detailed discussions on their applications in different contexts. We primarily focused on three distinct vehicular domains (i.e., automotive, aviation, and maritime) to provide a broader view of integrating multi-purpose prototyping solutions that fall within the scope of either digital twin or closely related technologies. As we move forward, emerging research should focus on solving more critical problems e.g.,—hybrid platforms, AI-enabled solutions, and platforms with greater integration and computational capabilities that can be effective. The survey presents a collection of research works addressing challenges in the vehicular industry, which, if used wisely, could assist the industry, organizations, and researchers in improving their systems' performance and usability to higher levels of operational readiness.

References

1. Liu, Mengnan, et al. 2021. Review of digital twin about concepts, technologies, and industrial applications. *Journal of Manufacturing Systems* 58: 346–361.
2. Wang, Wenjuan, et al. 2023. Digital twins in operation and maintenance (O&P). In *Digital twin technologies in transportation infrastructure management*, 179–203. Springer
3. Kabir, Md Rafiul, Bhagawat Baanav Yedla Ravi, and Sandip Ray. 2025. Digital twin technologies for vehicular prototyping: a survey. *IEEE Open Journal of Intelligent Transportation Systems*.
4. Kabir, Md Rafiul, et al. 2024. Automotive functional safety: scope, standards, and perspectives on practice. *IEEE Consumer Electronics Magazine* 14 (1): 10–25.
5. Wolf, Marilyn and Dimitrios Serpanos. 2017. Safety and security in cyber-physical systems and internet-of-things systems. *Proceedings of the IEEE* 106 (1): 9–20.
6. Kabir, Md Rafiul, Bhagawat Baanav Yedla Ravi, and Sandip Ray. 2023. A virtual prototyping platform for exploration of vehicular electronics. *IEEE Internet of Things Journal* 1–1. https://doi.org/10.1109/JIOT.2023.3267339.
7. Liao, Xishun, et al. 2023. Driver digital twin for online prediction of personalized lane change behavior. *IEEE Internet of Things Journal*.
8. Sakaguchi, T., A. Uno, and S. Tsugawa. 1997. An algorithm for merging control of vehicles on highways. In *Proceedings of the 1997 IEEE/RSJ international conference on intelligent robot and systems. Innovative robotics for real-world applications. IROS '97*, vol. 3, V15–V16. https://doi.org/10.1109/IROS.1997.656800.
9. Medina, Alejandro Ivan Morales, Nathan Van de Wouw, and Henk Nijmeijer. 2015. Automation of a T-intersection using virtual platoons of cooperative autonomous vehicles. In *2015 IEEE 18th international conference on intelligent transportation systems*, 1696–1701. IEEE.
10. Warwick, Graham. 2015. USAF selects lead programs for 'digital twin' initiative. *Aviat Week Space Technol.*

11. Liao, Min, Guillaume Renaud, and Yan Bombardier. 2020. Airframe digital twin technology adaptability assessment and technology demonstration. *Engineering Fracture Mechanics* 225: 106793.
12. Rudd, J. 2013. Airframe digital twin. In *Proceedings of the 27th symposium of the international committee on aeronautical fatigue and structural integrity, Jerusalem, Israel*, vol. 5, 15.
13. Ghosh, Sourabh, et al. 2022. Digital twin for electric energy systems: A new era of digitization. *IEEE Smart Grid*.
14. Digital Twin Market Size, Industry Report, Trend. 2023. Accessed 09 Jun 2024. Allied Market Research. https://www.alliedmarketresearch.com/digital-twin-market.
15. Digital Twins in Automotive Market. 2023. Accessed 09 Jun 2024. Allied Market Research. https://www.alliedmarketresearch.com/digital-twins-in-automotive-market.
16. Rawat, Danda B., and Chandra Bajracharya. 2017. *Vehicular cyber physical systems*, vol. 10, 978–3. Springer.
17. Lu, Yunlong, et al. 2020. Federated learning for data privacy preservation in vehicular cyber-physical systems. *IEEE Network* 34 (3): 50–56.
18. Jia, Dongyao, et al. 2015. A survey on platoon-based vehicular cyber-physical systems. *IEEE Communications Surveys & Tutorials* 18 (1): 263–284.
19. Mythily, M., Beaulah David, and J. Antony Vijay. 2024. Digital twin application in various sectors. In *Transforming industry using digital twin technology*. ed. Ashutosh Mishra, May El Barachi, and Manoj Kumar, 219–237. Cham: Springer Nature Switzerland. https://doi.org/10.1007/978-3-031-58523-4_11.
20. Mishra, Ashutosh, May El Barachi, and Manoj Kumar. 2024. *Transforming industry using digital twin technology*. Springer.
21. Mao, Runze, Yuanjiang Li, and Houxiang Zhang. 2023. Simulation method in automotive, aviation and maritime industries for digital twin: A brief survey. In *2023 IEEE 18th conference on industrial electronics and applications (ICIEA)*, 1442–1447. https://doi.org/10.1109/ICIEA58696.2023.10241843.
22. Sancarlos, Abel, et al. 2021. From ROM of electrochemistry to AI-based battery digital and hybrid twin. *Archives of Computational Methods in Engineering* 28: 979–1015.
23. Li, Luning, et al. 2022. Digital twin in aerospace industry: A gentle introduction. *IEEE Access* 10: 9543–9562. https://doi.org/10.1109/ACCESS.2021.3136458.
24. Rasheed, Adil, Omer San, and Trond Kvamsdal. 2020. Digital twin: Values, challenges and enablers from a modeling perspective. *IEEE Access* 8: 21980–22012.
25. Cubillo, Adrian, Suresh Perinpanayagam, and Manuel Esperon-Miguez. 2016. A review of physics-based models in prognostics: Application to gears and bearings of rotating machinery. *Advances in Mechanical Engineering* 8 (8): 1687814016664660.
26. Liu, Shimin, et al. 2021. Digital twin modeling method based on biomimicry for machining aerospace components. *Journal of Manufacturing Systems* 58: 180–195.
27. Côté, Isabelle M., and Karen L. Cheney. 2005. Choosing when to be a cleaner-fish mimic. *Nature* 433 (7023): 211–212.
28. Zhang, Cetengfei, et al. 2022. Dedicated adaptive particle swarm optimization algorithm for digital twin based control optimization of the plug-in hybrid vehicle. *IEEE Transactions on Transportation Electrification*.
29. Prisacaru, Alexandru, et al. 2021. Towards virtual twin for electronic packages in automotive applications. *Microelectronics Reliability* 122: 114134.
30. Li, Hao, et al. 2022. A detection and configuration method for welding completeness in the automotive body-in-white panel based on digital twin. *Scientific Reports* 12 (1): 7929.
31. Kumar, Sathish AP, et al. 2018. A novel digital twin-centric approach for driver intention prediction and traffic congestion avoidance. *Journal of Reliable Intelligent Environments* 4 (4): 199–209.

32. Rosenfeld, Avi, et al. 2015. Learning drivers' behavior to improve adaptive cruise control. *Journal of Intelligent Transportation Systems* 19 (1): 18–31.
33. Xiong, Minglan, et al. 2021. Digital twin–driven aero-engine intelligent predictive maintenance. *The International Journal of Advanced Manufacturing Technology* 114 (11): 3751–3761.
34. Pang, Yutian, Nan Xu, and Yongming Liu. 2019. Aircraft trajectory prediction using LSTM neural network with embedded convolutional layer. In *Proceedings of the annual conference of the PHM society*, vol. 11, 1–8. Scottsdale, AZ, USA: PHM Society.
35. Michael, Schultz, and Stefan Reitmann. 2018. Prediction of aircraft boarding time using LSTM network. In *2018 winter simulation conference (WSC)*, 2330–2341. IEEE.
36. Kosova, Furkan, and Hakki Ozgur Unver. 2023. A digital twin framework for aircraft hydraulic systems failure detection using machine learning techniques. *Proceedings of the Institution of Mechanical Engineers, Part C: Journal of Mechanical Engineering Science* 237 (7): 1563–1580.
37. Bouarfa, Soufiane, et al. 2020. Towards automated aircraft maintenance inspection. A use case of detecting aircraft dents using Mask R-CNN. In *AIAA Scitech 2020 forum*, 0389.
38. Yongbo, L.I., et al. 2020. Rotating machinery fault diagnosis based on convolutional neural network and infrared thermal imaging. *Chinese Journal of Aeronautics* 33 (2): 427–438.
39. Coraddu, Andrea, et al. 2019. Data-driven ship digital twin for estimating the speed loss caused by the marine fouling. *Ocean Engineering* 186: 106063.
40. Shen, Xingwang, et al. 2022. Digital twin-based scheduling method for marine equipment material transportation vehicles. In *2022 IEEE 18th international conference on automation science and engineering (CASE)*, 100–105. IEEE.
41. Zhang, Houxiang, et al. 2022. A digital twin of the research vessel gunnerus for lifecycle services: Outlining key technologies. *IEEE Robotics & Automation Magazine* 30 (3): 6–19.
42. Deng, Shutong, et al. 2023. A systematic review on the current research of digital twin in automotive application. *Internet of Things and Cyber-Physical Systems.*
43. Wang, Ziran, et al. 2022. Mobility digital twin: Concept, architecture, case study, and future challenges. *IEEE Internet of Things Journal* 9 (18): 17452–17467.
44. Niaz, Ashfaq, et al. 2021. Autonomous driving test method based on digital twin: A survey. In *2021 international conference on computing, electronic and electrical engineering (ICE Cube)*, 1–7. IEEE.
45. Hu, Zhongxu, et al. 2022. Review and perspectives on driver digital twin and its enabling technologies for intelligent vehicles. *IEEE Transactions on Intelligent Vehicles* 7 (3): 417–440.
46. Ma, Yunsheng, et al. 2024. Driver digital twin for online recognition of distracted driving behaviors. *IEEE Transactions on Intelligent Vehicles.*
47. Wang, Ziran, et al. 2020. Driver behavior modeling using game engine and real vehicle: A learning-based approach. *IEEE Transactions on Intelligent Vehicles* 5 (4): 738–749.
48. Wei, Chongfeng, et al. 2021. Human-like decision making and motion control for smooth and natural car following. *IEEE Transactions on Intelligent Vehicles* 8 (1): 263–274.
49. Canale, Massimo, Stefano Malan, and V. Murdocco. 2002. Personalization of ACC stop and go task based on human driver behaviour analysis. *IFAC Proceedings Volumes* 35 (1): 357–362.
50. Lefèvre, Stéphanie, et al. 2015. Driver models for personalised driving assistance. *Vehicle System Dynamics* 53 (12): 1705–1720.
51. Wang, Wenshuo, et al. 2018. A learning-based approach for lane departure warning systems with a personalized driver model. *IEEE Transactions on Vehicular Technology* 67 (10): 9145–9157.
52. Schnelle, Scott, et al. 2016. A driver steering model with personalized desired path generation. *IEEE Transactions on Systems, Man, and Cybernetics: Systems* 47 (1): 111–120.
53. Schnelle, Scott, et al. 2016. A personalizable driver steering model capable of predicting driver behaviors in vehicle collision avoidance maneuvers. *IEEE Transactions on Human-Machine Systems* 47 (5): 625–635.

54. Li, Lin, et al. 2016. Human dynamics based driver model for autonomous car. *IET Intelligent Transport Systems* 10 (8): 545–554.
55. Tselentis, Dimitrios I., and Eleonora Papadimitriou. 2023. Driver profile and driving pattern recognition for road safety assessment: Main challenges and future directions. *IEEE Open Journal of Intelligent Transportation Systems* 4: 83–100.
56. Yang, Kui, et al. 2022. Classification and evaluation of driving behavior safety levels: A driving simulation study. *IEEE Open Journal of Intelligent Transportation Systems* 3: 111–125.
57. Debada, Ezequiel, Laleh Makarem, and Denis Gillet. 2017. A virtual vehicle based coordination framework for autonomous vehicles in heterogeneous scenarios. In *2017 IEEE international conference on vehicular electronics and safety (ICVES)*, 51–56. https://doi.org/10.1109/ICVES.2017.7991900.
58. Leitao, J.M., A.A. Sousa, and F.N. Ferreira. 2000. Graphical control of autonomous, virtual vehicles. In *VTC2000-spring. 2000 IEEE 51st vehicular technology conference proceedings (cat. no. 00CH37026)*, vol. 1, 507–511. https://doi.org/10.1109/VETECS.2000.851509.
59. Egerstedt, Magnus, Xiaoming Hu, and Alexander Stotsky. 2001. Control of mobile platforms using a virtual vehicle approach. *IEEE Transactions on Automatic Control* 46 (11): 1777–1782.
60. Kyrkjebo, Erik, and Kristin Y. Pettersen. 2006. A virtual vehicle approach to output synchronization control. In *Proceedings of the 45th IEEE conference on decision and control*, 6016–6021. IEEE.
61. Deter, Dean, et al. 2021. Simulating the autonomous future: A look at virtual vehicle environments and how to validate simulation using public data sets. *IEEE Signal Processing Magazine* 38 (1): 111–121. https://doi.org/10.1109/MSP.2020.2984428.
62. Lei, T., et al. 2017. A cooperative route choice approach via virtual vehicle in IoV. *Vehicular Communications* 10: 116–124.
63. Liu, H.X., and W. Ma. 2009. A virtual vehicle probe model for time-dependent travel time estimation on signalized arterials. *Transportation Research Part C: Emerging Technologies* 17 (1): 11–26.
64. Tsanakas, N., J. Ekström, and J. Olstam. 2022. Generating virtual vehicle trajectories for the estimation of emissions and fuel consumption. *Transportation Research Part C: Emerging Technologies* 134: 103509.
65. Mecheri, S., and R. Lobjois. 2018. Steering control in a low-cost driving simulator: A case for the role of virtual vehicle cab. *Human Factors*.
66. Debada, E.G., and D. Gillet. 2018. Virtual vehicle-based cooperative maneuver planning for connected automated vehicles at single-lane roundabouts. In *IEEE intelligent transportation systems conference (ITSC)*. IEEE.
67. Xue, W., et al. 2019. An adaptive model predictive approach for automated vehicle control in fallback procedure based on virtual vehicle scheme. In *IEEE conference on connected vehicles*. IEEE.
68. Lin, J.T., F.K. Wang, and J.R. Young. 2004. Virtual vehicle in the connecting transport automated material-handling system (AMHS). *International Journal of Production Research*.
69. Ahn, Sunha, and Sharad Malik. 2014. Automated firmware testing using firmware-hardware interaction patterns. In *CODES+ISSS*, 25:1–25:10.
70. Kannavara, R., et al. 2015. Challenges and opportunities in concolic testing. In *Proceedings of the national aerospace electronics conference—Ohio innovation summit*.
71. Synopsys. 2019. Virtualizer. https://www.synopsys.com/verification/virtual-prototyping/virtualizer.html.
72. Gruyer, Dominique, and Mélanie. Grapinet, and Philippe De Souza. 2012. Modeling and validation of a new generic virtual optical sensor for ADAS prototyping. In *2012 IEEE intelligent vehicles symposium*, 969–974. IEEE.

73. Borg, Markus, et al. 2021. Digital twins are not monozygotic–cross-replicating adas testing in two industry-grade automotive simulators. In *2021 14th IEEE conference on software testing, verification and validation (ICST)*, 383–393. IEEE.
74. Menhour, Lghani, et al. 2017. An efficient model-free setting for longitudinal and lateral vehicle control: Validation through the interconnected Pro-SiVIC/RTMaps prototyping platform. *IEEE Transactions on Intelligent Transportation Systems* 19 (2): 461–475.
75. Mochizuki, Seiji, and Hirotaka, Hara. 2021. R-car virtual platform accelerates automotive software development for next-generation vehicles. Renesas Electronics. https://www.renesas.com/document/renesas-rcar-virtual-platform. Accessed 14 Jun 2024.
76. Strobl, Marius, et al. 2013. Towards automotive virtualization. In *2013 international conference on applied electronics*, 1–6. IEEE.
77. Safar, Mona, et al. 2019. Virtual verification and validation of automotive system. *Journal of Circuits, Systems and Computers* 28 (04): 1950071.
78. Lee, Chiyoung, Se-Won Kim, and Chuck Yoo. 2015. VADI: GPU virtualization for an automotive platform. *IEEE Transactions on Industrial Informatics* 12 (1): 277–290.
79. Hu, Xuemin, et al. 2018. Dynamic path planning for autonomous driving on various roads with avoidance of static and moving obstacles. *Mechanical Systems and Signal Processing* 100: 482–500. ISSN: 0888-3270. https://doi.org/10.1016/j.ymssp.2017.07.019. https://www.sciencedirect.com/science/article/pii/S0888327017303825.
80. Shiller, Zvi, Yu-Rwei Gwo, et al. 1991. Dynamic motion planning of autonomous vehicles. *IEEE Transactions on Robotics and Automation* 7 (2): 241–249.
81. Altché, Florent, Philip Polack, and Arnaud de La Fortelle. 2017. High-speed trajectory planning for autonomous vehicles using a simple dynamic model. In *2017 IEEE 20th international conference on intelligent transportation systems (ITSC)*, 1–7. IEEE.
82. Geisslinger, Maximilian, et al. 2023. Maximum acceptable risk as criterion for decision-making in autonomous vehicle trajectory planning. *IEEE Open Journal of Intelligent Transportation Systems*.
83. Wang, Ziran, Kyungtae Han, and Prashant Tiwari. 2021. Digital twin-assisted cooperative driving at non-signalized intersections. *IEEE Transactions on Intelligent Vehicles* 7 (2): 198–209.
84. Hu, Xuemin, et al. 2023. How simulation helps autonomous driving: A survey of sim2real, digital twins, and parallel intelligence. *IEEE Transactions on Intelligent Vehicles*.
85. Guo, Jiajie, et al. 2022. Survey on digital twins for internet of vehicles: Fundamentals, challenges, and opportunities. *Digital Communications and Networks*.
86. Bhatti, Ghanishtha, Harshit Mohan, and R. Raja Singh. 2021. Towards the future of smart electric vehicles: Digital twin technology. *Renewable and Sustainable Energy Reviews* 141: 110801.
87. Almeaibed, Sadeq, et al. 2021. Digital twin analysis to promote safety and security in autonomous vehicles. *IEEE Communications Standards Magazine* 5 (1): 40–46. https://doi.org/10.1109/MCOMSTD.011.2100004.
88. Thonhofer, Elvira, et al. 2023. Infrastructure-based digital twins for cooperative, connected, automated driving and smart road services. *IEEE Open Journal of Intelligent Transportation Systems* 4: 311–324.
89. Olgers, Tycho Joan, Anne Akke bij de Weg, and Jan Cornelis Ter Maaten. 2021. Serious games for improving technical skills in medicine: Scoping review. *JMIR Serious Games* 9 (1): e24093.
90. Bonyadi, Mohammad Reza, et al. 2016. Ahura: A heuristic-based racer for the open racing car simulator. *IEEE Transactions on Computational Intelligence and AI in Games* 9 (3): 290–304.
91. Samak, Tanmay Vilas, Chinmay Vilas Samak, and Ming Xie. 2021. Autodrive simulator: A simulator for scaled autonomous vehicle research and education. In *Proceedings of the 2021 2nd international conference on control, robotics and intelligent system*, 1–5.

92. Dosovitskiy, Alexey, et al. 2027. CARLA: An open urban driving simulator. In *Conference on robot learning*, 1–16. PMLR.
93. Owoputi, Richard, Md Rafiul Kabir, and Sandip Ray. 2023. IVE: An immersive virtual environment for automotive security exploration. *Immersive Learning Research-Academic* 468–480.
94. Zeng, Haibo, et al. 2006. Design space exploration of automotive platforms in metropolis. *SAE Transactions* 844–856.
95. Behrisch, Michael, et al. 2011. SUMO–simulation of urban mobility: An overview. In *Proceedings of SIMUL 2011, the third international conference on advances in system simulation*. ThinkMind.
96. Wymann, Bernhard, et al. 2000. Torcs, the open racing car simulator. *Software available at http://torcs. sourceforge. net* 4 (6): 2.
97. Tomandl, Andreas, et al. 2014. VANETsim: An open source simulator for security and privacy concepts in VANETs. In *2014 International conference on high performance computing and simulation (HPCS)*, 543–550. IEEE.
98. Kabir, Md Rafiul, and Sandip Ray. 2024. ViSE: Digital twin exploration for automotive functional safety and cybersecurity. *Journal of Hardware and Systems Security* 1–12.
99. Kabir, Md Rafiul, and Sandip Ray. 2023. Virtualization for automotive safety and security exploration. In *2023 IEEE 16th Dallas circuits and systems conference (DCAS)*, 1–4. IEEE.
100. Kabir, Md Rafiul, Neha Mishra, and Sandip Ray. 2021. Vive: Virtualization of vehicular electronics for system-level exploration. In *2021 IEEE international intelligent transportation systems conference (ITSC)*, 3307–3312. IEEE.
101. Kabir, Md Rafiul, Bhagawat Baanav Yedla Ravi, and Sandip Ray. 2023. A virtual prototyping platform for exploration of vehicular electronics. *IEEE Internet of Things Journal* 10 (18): 16144–16155.
102. Fellendorf, Martin, and Peter Vortisch. 2010. Microscopic traffic flow simulator VISSIM. In *Fundamentals of traffic simulation*, 63–93.
103. Li, Luning, et al. 2021. Digital twin in aerospace industry: A gentle introduction. *IEEE Access* 10: 9543–9562.
104. Xiong, Minglan, and Huawei Wang. 2022. Digital twin applications in aviation industry: A review. *The International Journal of Advanced Manufacturing Technology* 121 (9): 5677–5692.
105. Qiu, Z., and X. Wang. 2012. Comprehensive design ideas of modern aircraft structure. In *Fundamentals of aircraft structural strength analysis and design*, 5–7. Beijing, China: University of Aeronautics and Astronautics Press.
106. Du, Y. 2014. Research on probabilistic damage tolerance analysis system and key technologies. PhD thesis, Northwestern Polytechnical University, Xi'an, Shanxi, China, 1–2.
107. Lee, H., H. Cho, and S. Park. 2012. Review of the F-16 individual aircraft tracking program. *Journal of Aircraft* 49 (5): 1398–1405.
108. Bachelor, G., et al. 2019. Model-based design of complex aeronautical systems through digital twin and thread concepts. *IEEE Systems Journal* 14 (2): 1568–1579.
109. Tadeja, S., P. Seshadri, and P. Kristensson. 2020. AeroVR: An immersive visualisation system for aerospace design and digital twinning in virtual reality. *The Aeronautical Journal* 124 (1280): 615–1635.
110. Demartini, M., et al. 2019. Closed-loop manufacturing for aerospace industry: An integrated PLM-MOM solution to support the wing box assembly process. In *IFIP international conference on advances in production management systems*, 423–430. Cham: Springer.
111. Cai, H., W. Zhang, and Z. Zhu. 2019. Quality management and analysis of aircraft final assembly based on digital twin. In *2019 11th international conference on intelligent human-machine systems and cybernetics (IHMSC)*, vol. 1, 202–205. IEEE.
112. Cai, Hongxia, Jiamin Zhu, and Wei Zhang. 2021. Quality deviation control for aircraft using digital twin. *Journal of Computing and Information Science in Engineering* 21 (3): 031008.

113. Guo, F., et al. 2018. Working mode in aircraft manufacturing based on digital coordination model. *International Journal of Advanced Technology* 98 (5): 1547–1571.
114. Ezhilarasu, Cordelia Mattuvarkuzhali, Zakwan Skaf, and Ian K. Jennions. 2019. Understanding the role of a digital twin in integrated vehicle health management (IVHM). In *2019 IEEE international conference on systems, man and cybernetics (SMC)*, 1484–1491. https://doi.org/10.1109/SMC.2019.8914244.
115. Redding, Louis. 2011. An introduction to integrated vehicle health management. *Integrated Vehicle Health Management: Perspectives on an Emerging Field*, 17.
116. Li, Chenzhao, et al. 2017. Dynamic Bayesian network for aircraft wing health monitoring digital twin. *Aiaa Journal* 55 (3): 930–941.
117. Zakrajsek, Andrew J., and Shankar Mall. 2017. The development and use of a digital twin model for tire touchdown health monitoring. *58th AIAA/ASCE/AHS/ASC structures, structural dynamics, and materials conference*, 0863.
118. Trusova, K.V. 2022. Synthesis and analysis of avionics functions digital twins using machine learning classification algorithms. In *Information technologies: proceedings of the YETI 2021*. Springer.
119. Seshadri, Banavara R., and Thiagarajan Krishnamurthy. 2017. Structural health management of damaged aircraft structures using digital twin concept. In *25th aiaa/ahs adaptive structures conference*, 1675.
120. Lai, Xiaonan, et al. 2023. Digital twin-based structural health monitoring by combining measurement and computational data: An aircraft wing example. *Journal of Manufacturing Systems* 69: 76–90.
121. Zaccaria, Valentina, et al. 2018. Fleet monitoring and diagnostics framework based on digital twin of aero-engines. In *Turbo expo: power for land, sea, and air*, vol. 51128, V006T05A021. American Society of Mechanical Engineers.
122. Hartwell, A., et al. 2024. Distributed digital twins for health monitoring: Resource constrained aero-engine fleet management. *The Aeronautical Journal* 1–20.
123. Sadeghi, Alireza, et al. 2024. Digital twins for condition and fleet monitoring of aircraft: Towards more-intelligent electrified aviation systems. *IEEE Access*.
124. Liu, Zheng, Norbert Meyendorf, and Nezih Mrad. 2018. The role of data fusion in predictive maintenance using digital twin. In *AIP conference proceedings*, vol. 1949, 1. AIP Publishing.
125. Utzig, Sebastian, et al. 2019. Augmented reality for remote collaboration in aircraft maintenance tasks. In *2019 IEEE aerospace conference*, 1–10. IEEE.
126. Pickup, O. 2018. What is a digital twin and how does it keep Rolls-Royce machines safe. *The Telegraph*.
127. Chowdhury, Shafayat Hasan, Fakhre Ali, and Ian K Jennions. 2019. A methodology for the experimental validation of an aircraft ECS digital twin targeting system level diagnostics. In *Annual conference of the PHM society*, vol. 11, no. 1, 21–26.
128. Lv, Zhihan, et al. 2021a. Beyond 5G for digital twins of UAVs. *Computer Networks* 197: 108366.
129. Lv, Zhihan, et al. 2021b. Digital twins in unmanned aerial vehicles for rapid medical resource delivery in epidemics. *IEEE Transactions on Intelligent Transportation Systems* 23 (12): 25106–25114.
130. Lei, L., et al. 2020. Toward intelligent cooperation of UAV swarms: When machine learning meets digital twin. *IEEE Network*.
131. Shen, G., et al. 2021. Deep reinforcement learning for flocking motion of multi-UAV systems: Learn from a digital twin. *IEEE Internet of Things Journal*.
132. Shen, Gaoqing, et al. 2023. Multi-UAV cooperative search based on reinforcement learning with a digital twin driven training framework. *IEEE Transactions on Vehicular Technology*.

133. Li, Siyuan, et al. 2022. When digital twin meets deep reinforcement learning in multi-UAV path planning. *Proceedings of the 5th international ACM mobicom workshop on drone assisted wireless communications for 5G and beyond*, 61–66.
134. Tang, Xin, et al. 2023. Digital-twin-assisted task assignment in multi-UAV systems: A deep reinforcement learning approach. *IEEE Internet of Things Journal* 10 (17): 15362–15375.
135. Yang, Yuanlin, Wei Meng, and Shiquan Zhu. 2020. A digital twin simulation platform for multi-rotor UAV. In *2020 7th international conference on information, cybernetics, and computational social systems (ICCSS)*, 591–596. IEEE.
136. Meng, Wei, et al. 2023. DTUAV: A novel cloud–based digital twin system for unmanned aerial vehicles. *Simulation* 99 (1): 69–87.
137. Hazarika, Bishmita, et al. 2023. RADiT: Resource allocation in digital twin-driven UAV-aided internet of vehicle networks. *IEEE Journal on Selected Areas in Communications.*
138. Aláez, Daniel, et al. 2023, VTOL UAV digital twin for take-off, hovering and landing in different wind conditions. *Simulation Modelling Practice and Theory* 123: 102703.
139. Li, Bin, et al. 2023. Adaptive digital twin for UAV-assisted integrated sensing, communication, and computation networks. *IEEE Transactions on Green Communications and Networking.*
140. Soliman, Abdulrahman, et al. 2023. AI-based UAV navigation framework with digital twin technology for mobile target visitation. *Engineering Applications of Artificial Intelligence* 123: 106318.
141. Chen, Shazhou, et al. 2020. A warehouse management system with uav based on digital twin and 5g technologies. In *2020 7th international conference on information, cybernetics, and computational social systems (ICCSS)*, 864–869. IEEE.
142. Kandaz, M., et al. 2024. Simulation-based digital twin of a high-speed turbomachine (fan) used in avionics cooling. *Engineering Modelling, Analysis and Simulation.*
143. Fraser, Benjamin, et al. 2021. Enhancing the security of unmanned aerial systems using digital-twin technology and intrusion detection. In *2021 IEEE/AIAA 40th digital avionics systems conference (DASC)*, 1–10. IEEE.
144. Kuleshov, Y.A., K. Nagpal, K. Ucpinar, et al. 2024. Cyber attacks on avionics networks in digital twin environment: Detection and defense. In *AIAA SCITECH 2024 forum.*
145. Mao, Runze, YuanJiang Li, and Houxiang Zhang. 2023. Digital twin-based research in the maritime industry: A brief survey. In *IECON 2023-49th annual conference of the IEEE industrial electronics society*, 1–6. IEEE.
146. Fonseca, Ícaro Aragão, and Henrique Murilo Gaspar. 2021. Challenges when creating a cohesive digital twin ship: A data modelling perspective. *Ship Technology Research* 68 (2): 70–83.
147. Mombiela, Dalia Casanova, and Mehdi Zadeh. 2021. Integrated design and control approach for marine power systems based on operational data; digital twin to design. In *2021 IEEE transportation electrification conference and expo (ITEC)*, 520–527. IEEE.
148. Menges, Daniel, Andreas Von Brandis, and Adil Rasheed. 2024. Digital twin for autonomous surface vessels for safe maritime navigation. arXiv:2401.04032.
149. Kinaci, Omer Kemal. 2023. Ship digital twin architecture for optimizing sailing automation. *Ocean Engineering* 275: 114128.
150. Perabo, Florian, et al. 2020. Digital twin modelling of ship power and propulsion systems: Application of the open simulation platform (osp). In *2020 IEEE 29th international symposium on industrial electronics (ISIE)*, 1265–1270. IEEE.
151. Arrichiello, Vincenzo, and Paola Gualeni. 2020. Systems engineering and digital twin: A vision for the future of cruise ships design, production and operations. *International Journal on Interactive Design and Manufacturing (IJIDeM)* 14: 115–122.
152. Hasan, Agus, et al. 2023. Predictive digital twins for autonomous surface vessels. *Ocean Engineering* 288: 116046.

153. Liu, Jun, et al. 2021. Security in IoT-enabled digital twins of maritime transportation systems. *IEEE Transactions on Intelligent Transportation Systems* 24 (2): 2359–2367.
154. Kaklis, Dimitrios, et al. 2023. Enabling digital twins in the maritime sector through the lens of AI and industry 4.0. *International Journal of Information Management Data Insights* 3 (2): 100178.
155. Pedersen, Tom Arne, et al. 2020. Towards simulation-based verification of autonomous navigation systems. *Safety Science* 129: 104799.
156. Wang, Zhen, and Bin Lin. 2022. A digital twin enabled maritime networking architecture. In *2022 IEEE 95th vehicular technology conference: (VTC2022-spring)*, 1–5. IEEE.
157. Bhagavathi, Ravitej, D. Kwame Minde Kufoalor, and Agus Hasan. 2023. Digital twin-driven fault diagnosis for autonomous surface vehicles. *IEEE Access.*
158. Hasan, Agus, et al. 2024. Leveraging digital twins for fault diagnosis in autonomous ships. *Ocean Engineering* 292: 116546.
159. Gao, Shuxian, et al. 2021. Digital twin escort system for underwater unmanned vehicle. In *International conference on autonomous unmanned systems*, 1082–1089. Springer.
160. Martynova, Liubov, Ivan Pashkevich, and Valentina Bykova, 2024. Development of a digital twin of an autonomous underwater vehicle to assess the effectiveness of searching for bottom objects. In *2024 international Russian smart industry conference (SmartIndustryCon)*, 243–248. IEEE.
161. Hu, Shanshan, et al. 2023. Construction of a digital twin system for the blended-wing-body underwater glider. *Ocean Engineering* 270: 113610.
162. Wen, Jiabao, et al. 2022. Behavior-based formation control digital twin for multi-AUG in edge computing. *IEEE Transactions on Network Science and Engineering.*
163. Yang, Ming, et al. 2023. Digital twin-driven industrialization development of underwater gliders. *IEEE Transactions on Industrial Informatics.*
164. Kamdjou, Hugues M., et al. 2024. Resource-constrained extended reality operated with digital twin in industrial internet of things. *IEEE Open Journal of the Communications Society* 5: 928–950. https://doi.org/10.1109/OJCOMS.2024.3356508.

Digital Twins in Internet of Medical Things 4

From digital patients to smart hospitals, digital twins in the Internet of Medical Things.

4.1 Introduction

The Internet of Things (IoT) has seamlessly integrated into the fabric of daily life, connecting an ever-expanding array of devices and systems—from smartphones and smart homes to intelligent building management systems and health-monitoring wearables [1]. This integration extends significantly into the healthcare sector, where IoT technologies have revolutionized the way health systems operate. By embedding sensors and devices within the healthcare framework, IoT has enabled a more streamlined workflow, accelerated access to medical records, and enhanced the accuracy and reliability of data collected from diverse sources [2]. IoT's capacity to facilitate seamless data sharing among healthcare providers has also been crucial in combating global health crises, such as pandemics. By providing real-time data integration and communication, IoT systems help in tracking disease spread, managing resources, and implementing targeted responses more efficiently. Additionally, the technology plays a pivotal role in improving patient care by enabling the continuous monitoring of health conditions through wearable devices that communicate directly with healthcare providers. This direct line of communication ensures that any significant changes in a patient's condition are promptly addressed, potentially saving lives by allowing for immediate medical intervention [3].

In order to provide proactive patient care, medical professionals and engineers are moving towards using the digital twin (DT) technology to increase patients' quality of life. By fusing AI and ML functionalities with different IoT elements, e.g., cloud apps, DTs can use real-time data to make accurate predictions about people's health outcomes [4]. In its simplest form, a digital twin is indeed just a virtual model of a physical object. However, this basic definition

M. R. Kabir and S. Ray, *Digital Twins for Distributed IoT Applications*, Synthesis Lectures on Engineering, Science, and Technology, https://doi.org/10.1007/978-3-032-18938-7_4

barely scratches the surface of what DT technology entails and the complex functionalities it encompasses. Unlike a straightforward 3D model, which is static and merely represents the physical appearance of an object, a DT is dynamic and interactive. It integrates real-time data continuously updated from its physical counterpart, allowing it to simulate behavior and process in a real-world context.

The true power of DT lies in its versatility and depth. For instance, in the healthcare sector, a DT could be used to model complex biological systems such as the human heart. This isn't just a structural replica but a functional one that can mimic the actual working conditions of the heart under various scenarios. By applying predictive analytics and machine learning algorithms, the DT can forecast potential health issues before they occur by analyzing deviations from normal operation patterns. Moreover, a DT could represent broader systems such as a hospital's operations or even an entire population's health dynamics. These models can simulate public health responses and predict the outcomes of epidemics, helping in strategic planning and resource allocation. A DT is not only about visual representation but also about behavior and function. It is an evolving model that learns from multiple data streams, adapts to changes, and provides predictive insights. This capability makes DTs invaluable across various sectors, including manufacturing, automotive, and urban planning, where they contribute to enhanced efficiency, innovation, and decision-making.

Amid these technological advances, we present a comprehensive overview of the integration of DT technology within the healthcare IoT ecosystem, emphasizing its critical role in advancing personalized medicine and addressing the significant security challenges that are inherent in healthcare IoT systems. Concentrating exclusively on the field of smart healthcare systems with the IoT infrastructure, we explore the application of DTs in various facets of healthcare. Our investigation includes digital patient modeling, which utilizes DTs to create highly accurate and dynamic patient profiles that help in tailoring medical treatments more effectively. We also examine smart hospital lifecycle management, where DTs optimize hospital infrastructure and patient flow. Additionally, we discuss DT-based surgical applications, from preoperative and intraoperative planning to real-time surgical guidance.

Beyond clinical applications, we explore DT integration with medical devices, enhancing functionality, predictive maintenance, and safety. Additionally, we explore the impact of DTs in the pharmaceutical industry. Here, DTs are instrumental in streamlining drug development processes, from simulation of clinical trials to monitoring the manufacturing and distribution chains, thus ensuring higher efficiency and compliance with regulatory standards. Finally, we address security and privacy challenges in IoMT, an emerging yet critical area of research. By demonstrating these diverse applications, we highlight the transformative potential of digital twins in modern healthcare, showcasing how they can lead to more efficient, safe, and personalized medical services. This chapter not only underscores the benefits but also discusses the technological and ethical challenges involved in integrating DTs into the healthcare sector, proposing pathways for future research and development. Figure 4.1 shows a taxonomy of digital usage in the smart healthcare domain.

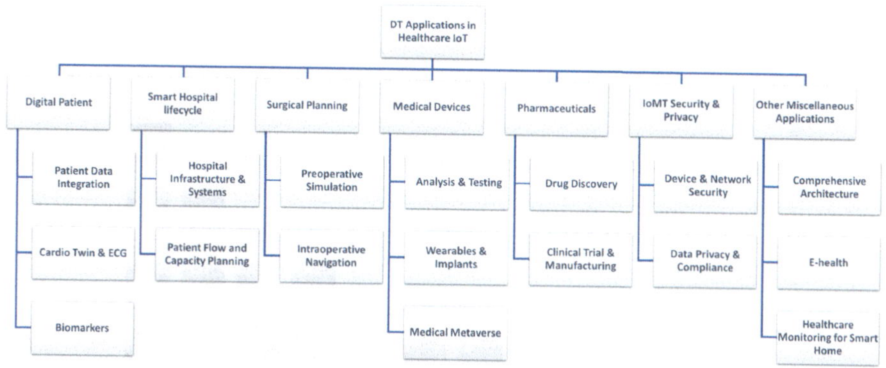

Fig. 4.1 A taxonomy of digital twins in smart healthcare domain

4.2 Background and Concepts

4.2.1 Digital Twin Concept in Healthcare

Some studies have used the terms "Medical Digital Twin" [5, 6] and "Health Digital Twin" [7, 8] to represent simulation modeling focused on specific components of the healthcare sector. Another term "Human Digital Twin (HDT)" has been used by many studies that extend the DT technology beyond its applications in healthcare [9, 10] to a wide range of engineering, industrial, and social domains [11–15]. An HDT is a dynamic digital representation of an individual human being, capturing in real-time the complex interactions of biological, psychological, and social factors.

4.2.2 Healthcare DTs Market Trends

The healthcare DT market is poised for significant growth, with an estimated market size of USD 902.59 million in 2024, projected to grow at a CAGR of 25.9% through 2030 [16]. DTs are revolutionizing the healthcare sector by providing real-time, integrated, and interactive approaches for data capture, simulation, and feedback, enabling improvements in efficiency, cost optimization, and future demand anticipation. The rising adoption of sensors, medical records, wearables, and mobile applications is driving this transformation, as these tools produce valuable data for simulating pharmaceuticals and medical equipment. Recent investments, such as Accenture's partnership with Virtonomy, highlight the push to accelerate innovation in life-saving medical devices using DT technology. Furthermore, funding initiatives, including over USD 6 million in grants from the U.S. National Science Foundation in collaboration with the NIH and FDA, underscore the growing momentum in

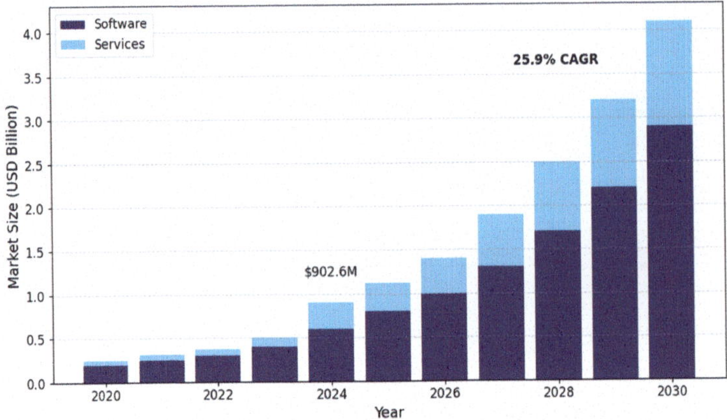

Fig. 4.2 Healthcare digital twins market size by component, 2020–2030 (USD Billion)

DT research for healthcare and biomedical applications, promising a new era of personalized medicine and medical innovation. Figure 4.2 illustrates the healthcare DT market size by component from 2020 to 2030.

Building on this momentum, the healthcare DT market is witnessing rapid growth across key segments. The providers segment leads with the largest market share at 36.2% in 2024 (see Fig. 4.3), as DTs are increasingly utilized in healthcare facilities to optimize staff man-

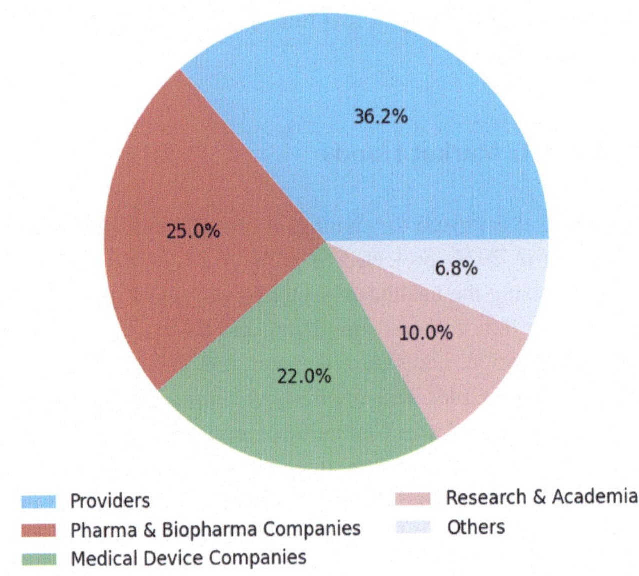

Fig. 4.3 Digital twins end-use market share in the healthcare industry

agement and detect resource shortages, such as bed availability, enabling more efficient decision-making and cost reduction [16]. For example, IBM's Process Mining software generates DTs to simulate hospital operations and identify potential challenges. Partnerships like the one between Strasys and Silico in December 2022 aim to deliver DTs for entire hospitals and healthcare systems, enhancing decision intelligence. Meanwhile, the pharma and biopharma segment is expected to grow at the fastest CAGR, driven by the demand for cost-effective drug development and personalized medicine. DTs are being used to simulate patient responses, optimize dosages, and develop targeted therapies, thereby accelerating the adoption of this technology across the industry.

4.2.3 Healthcare IoT

The concept of Healthcare IoT, or the Internet of Medical Things (IoMT), refers to the integration of connected devices in the medical and healthcare sectors to improve the efficiency and effectiveness of medical services. In simpler terms, this involves medical devices capable of sending data over a network autonomously, without the need for direct human-to-human or human-to-computer interaction [17]. Many studies explored and initiated many aspects of IoT in healthcare in terms of architecture [18], implementation of an IoT based In-hospital healthcare [19], concepts [20], secure medical model [21, 22], etc. Here are some key aspects and benefits of Healthcare IoT:

- Remote Monitoring: One of the most significant advantages of Healthcare IoT is the ability to monitor patients remotely. Devices can collect data on vital signs like heart rate, blood pressure, glucose levels, and more [23]. This data is sent to healthcare providers in real-time, enabling them to monitor patients' conditions without needing them to be physically present in a medical facility [24]. This is particularly beneficial for managing chronic conditions and elderly care.
- Data Collection and Analysis: IoT devices gather vast amounts of health data, which can be analyzed using AI and machine learning algorithms to provide insights into patient health trends, predict health episodes, and optimize treatment plans [25, 26]. This data-driven approach can lead to more personalized and timely healthcare.
- Improved Patient Engagement and Compliance: Devices connected through the IoT can provide patients with regular updates and reminders about their health management tasks, such as taking medication, scheduling appointments, or exercising. This can help increase patient engagement with their health regimes and improve compliance with treatment plans.
- Enhanced Medical Device Management: IoT can also streamline the management of medical devices within healthcare facilities. For example, tracking devices can monitor the location and usage of medical equipment, helping to optimize resource allocation and reduce operational costs.

- Advanced Diagnostics and Treatment: IoT technologies enable the development of new diagnostic and treatment technologies. For instance, smart inhalers for asthma patients can track usage patterns and environmental triggers for asthma attacks, potentially leading to more effective management of the condition.
- Security and Privacy Concerns: Despite its benefits, Healthcare IoT raises significant security and privacy concerns, as medical data is highly sensitive and protecting it against cyber threats is crucial [27]. This requires robust encryption methods, secure data transmission protocols, and compliance with health data protection regulations.

Overall, Healthcare IoT holds the potential to revolutionize healthcare by making it more proactive, patient-centered, and efficient. However, it also necessitates careful consideration of privacy, security, and technical integration aspects to fully leverage its capabilities.

4.2.4 Related Surveys and Reviews

The scholarly interest in the integration of DTs within the smart healthcare sector has been extensively documented through various surveys and review papers. The majority of this research concentrates on replicating human organs, aiming to advance personalized medicine and improve treatment outcomes, while others focus on the application of hospital operations and medical devices. Table 4.1 here summarizes the existing literature reviews.

4.3 Applications in Smart Healthcare

4.3.1 Digital Patient

The advancement of IoT infrastructure within the healthcare sector has significantly enhanced the use of DTs, especially in intelligent human body monitoring and the maintenance of medical devices. While the broader medical field increasingly adopts this technology to create virtual representations of the human body i.e., "human digital twin (HDT)" or "digital patients," our discussion will specifically focus on those aspects of digital patient activities that are integrated with IoT components. This targeted exploration helps to understand how IoT connectivity can optimize health monitoring and device maintenance, thereby improving patient care and operational efficiency in healthcare settings. Figure 4.4 shows the conceptual overview of a DT of a heart model as an illustrative example.

Patient Data Integration

Vats et al. [9] developed an IoT-based framework using HDT technology and machine learning to predict cardiovascular disease early, utilizing data from wearable sensors and a

Table 4.1 Summary of related surveys and reviews on DTs in healthcare

References	Focus
Katsoulakis et al. [28]	Current applications, consortium research centers reviews, and emerging research opportunities
Sun et al. [29]	Application examples of DT in many fields of medicine
Elkefi et al. [30]	Application for improving healthcare management and discusses associated challenges
Coorey et al. [8]	Challenges and applications of DT for cardiovascular disease
Erol et al. [31]	Ongoing and planned DT studies in healthcare, highlighting future developments
Okegbile et al. [10]	Architectural framework, design requirements, and applications of HDT with future directions
Chen et al. [32]	Networking architecture and key technologies for HDT in personalized healthcare applications
Haleem et al. [27]	Various features, services, and tools of DT for healthcare
Bjelland et al. [33]	Concepts for arthroscopic knee surgery using patient-specific information and methods for soft tissue simulation
Narigina et al. [34]	Highlighting development, methodologies, trends, challenges, and potential to improve patient care and healthcare innovation
Kabir et al. [35]	Focused solely on the growing field of smart healthcare systems powered by IoT infrastructure
Xames et al. [36]	Using the PRISMA approach, identified research trends, gaps, and realization challenges

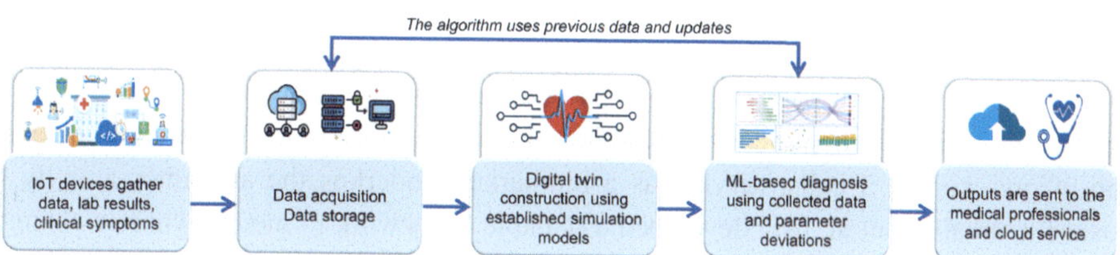

Fig. 4.4 Overview of a digital twin process of a patient organ: illustrative example for heart model

comprehensive dataset of 70,000 patient records. This approach achieved an impressive prediction accuracy of 88.91% with XG Boost, enhancing the integration of patients and doctors and improving healthcare operations and services. Mourtzis et al. [37] proposed a conceptual framework that leverages information and communication technologies (ICT), augmented reality (AR), and artificial intelligence (AI) to enhance the visualization, data acquisition, and analysis of MRI scans for cancer diagnosis and prediction. The paper also discusses the preliminary development of this framework and explores potential future advancements in light of the challenges encountered.

Cardio Twin and ECG

The majority of DT studies focusing on human organs have concentrated on the cardiology domain. Utilizing real-time ECG data, DTs of the heart allow for dynamic simulations and analysis, enhancing the accuracy of cardiac health assessments and the prediction of potential disorders. Comprehensive reviews (e.g., [8, 38] offer an extensive outlook on initiatives aimed at overcoming existing challenges and fully harnessing the potential of DTs to address cardiovascular disease. Elayan et al. [1] proposed and implemented an intelligent, context-aware healthcare system using a DT framework, enhancing digital healthcare by integrating real-time data for dynamic patient monitoring and operation improvement. The system includes an ML-based electrocardiogram (ECG) classifier that accurately diagnoses heart conditions, demonstrating how DT, combined with advanced neural network algorithms, can revolutionize continuous health monitoring and disease detection. Martinez et al. [39] introduced a Cardio Twin architecture, an edge-based system for ischemic heart disease detection using a convolutional neural network (CNN) that classifies conditions from electrocardiograms with 85.77% accuracy. Trained on data from 200 individuals, it highlights the feasibility of DT in demanding healthcare applications.

Biomarkers

Biomarkers are critical for creating personalized health profiles and predictive models. DTs using biomarker data can simulate disease progression and response to treatment at an individual level. Susilo et al. [40] developed a novel workflow to create DTs for each patient, forming a virtual population that represented variability in biological, pharmacological, and tumor-related parameters from the phase I trial, which predicted that increased mosunetuzumab exposure increases the proportion of patients with significant tumor reduction. They also identified that patients with indolent non-Hodgkin's lymphoma (NHL) showed increased sensitivity to treatment, with model simulations suggesting that intratumor expansion of pre-existing T-cells, serving as a biomarker, underlies the antitumor activity of mosunetuzumab. Li et al. [41] developed a scalable framework to model dynamic changes in DTs on cellulome- and genome-wide scales, which prioritizes upstream regulator (UR) genes for biomarker and drug discovery. Their approach, tested on single-cell and bulk-profiling data from seasonal allergic rhinitis (SAR) and other inflammatory diseases, successfully ranked URs based on predicted effects on downstream target cells, demonstrating a tractable method for UR prioritization.

4.3.2 Smart Hospital Lifecycle

The increasing complexity of hospital operations necessitates advanced digital solutions that enhance clinical decision-making, optimize resource allocation, and improve patient care. Traditional hospital management relies on fragmented data sources and isolated dig-

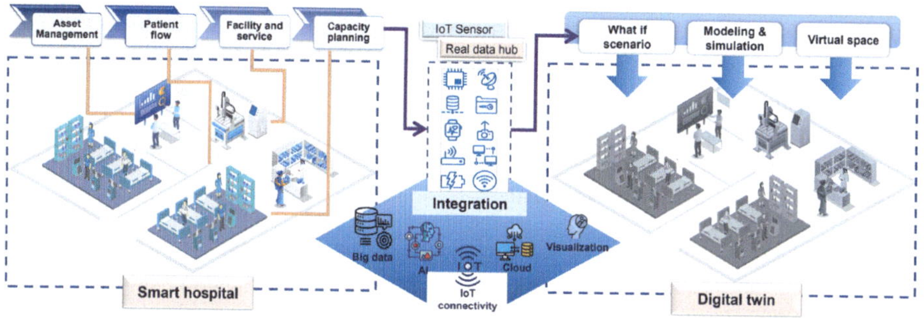

Fig. 4.5 Overview of a digital twin for smart hospital lifecycle

ital tools, limiting the ability to coordinate real-time responses and predictive analytics [42]. DTs address this by creating virtual replicas of hospital environments and integrating real-time data from IoT devices and medical records. While DTs have been widely used in manufacturing, their healthcare applications remain limited to isolated processes like surgical planning and equipment monitoring [43]. As hospitals adopt AI-driven, predictive models, DTs offer a pathway to smarter, more efficient healthcare systems. Figure 4.5 shows an overview of DT for the smart hospital lifecycle.

Hospital Infrastructure and Systems

The overall hospital infrastructure is pivotal in ensuring efficient and patient-centered care within smart healthcare environments. By integrating advanced IoT devices, DT technology, and interconnected systems, modern hospitals are transforming their infrastructure to enable seamless patient pathways, real-time monitoring, and personalized treatment strategies. HospiT'Win [44] is a framework that uses IoT, body sensor networks (BSN), modeling, simulation, and AI to create a DT—a virtual replica of a hospital—to monitor patient pathways, predict outcomes, and improve care delivery. Synchronizing real-time patient data with the DT, helps healthcare providers anticipate unexpected events, minimize their impact, and ensure timely, high-quality care. Karakra et al. [45] developed a DT model of hospital services using discrete event simulation (DES) integrated with healthcare information systems and IoT devices. This model enables real-time assessment of service efficiency and evaluates changes in healthcare delivery systems without disrupting hospital operations. For smart hospitals, Han et al. [46] proposed a conceptual DT framework to integrate real-time data from multiple sources, addressing both clinical and non-clinical operational needs. The pilot platform tested in a Shanghai municipal hospital demonstrated the potential to enhance digitization, automation, and intelligent control of hospital operations. Another interesting work by Aluvalu et al. [47] proposed a model for Emergency Room Service (ERS) using DTs, IoT devices, and sensors to fast-track patient treatment, reduce the length of stay, and assess risk factors by leveraging digital health records and biometric authentication. The model incorporates face recognition, cloud-based data access, and communication systems to notify families and experts, achieving an 80% success rate for an intelligent expert system to handle anonymous patients effectively.

Patient Flow and Capacity Planning

Gorelova et al. [48] conducted simulations of patient flow in two assisted reproduction clinics using discrete event simulation techniques to evaluate the impact of smart health monitoring systems (SHMSs). By analyzing multiple scenarios, they identified bottlenecks, optimized staff allocation, and proposed the most efficient SHMS implementation. Their findings demonstrated reduced workloads for receptionists, ultrasound and blood collection areas, and physicians, improving overall clinic efficiency. Another recent work et al. [49] introduced a novel framework for dynamic task offloading in healthcare that integrates DT and social health determinants to enhance patient care and system performance. They demonstrate notable efficiency gains, such as a 20% latency reduction and over 50% power savings, when scaling to 30 MEC nodes, thus bridging significant research gaps and advancing healthcare technology integration.

4.3.3 Surgical Planning

DTs in surgical planning involve using virtual models of patient anatomy to plan and simulate surgeries, enhancing precision and outcomes. This application impacts the operational and clinical aspects of hospital management.

Preoperative Simulation

Bjelland et al. [33] conducted a systematic review using the PRISMA protocol to summarize and analyze the literature on designing an arthroscopic DT with patient-specific information and methods for interactive surgical soft tissue simulation. They proposed a novel macro-level conceptual system that integrates patient-specific images, diagnostic data, intraoperative sensor data, and surgical practice inputs, aiming to enhance surgical skills training, and preoperative planning, and create a database of virtual surgeries. Kleinbeck et al. [50] developed a neural rendering-based method to create immersive DTs of complex medical environments from casual video capture, enabling spatial planning of surgical scenarios in virtual reality. Their evaluation showed that this approach significantly increased perceived utility and presence, with higher exploratory behavior, indicating its potential for effective spatial planning without requiring expert knowledge or specialized hardware. Tai et al. [51] created a DT-enabled IoMT system for telemedical simulation, integrating mixed reality (MR), 5G cloud computing, and a generative adversarial network (GAN) to achieve remote lung cancer implementation. Their system demonstrated high accuracy in lung cancer and pulmonary embolism prediction and enhanced remote surgical implementation by projecting real-time operating room perspectives to the surgeon, significantly improving clinical data processing and deep learning analytics for DT-based surgical procedures.

Intraoperative Navigation

The work of Tai et al. [51] mentioned above is also relevant here as it helps in real-time surgical implementation by projecting live operating room perspectives to the surgeon. In the case of remote surgery, Laaki et al. [52] developed a novel DT prototype to analyze communication requirements, focusing on low latency, high security, and reliability, using a robotic arm and HTC Vive VR system connected over a 4G mobile network. Despite limitations in VR capabilities and robot feedback, their research provided insights into communication establishment and cybersecurity technologies necessary for DT architecture development, highlighting the need for cross-disciplinary development and integration challenges in realizing Industry 4.0 concepts. A virtual TAVR planning process was performed by Obaid et al. [53], using patient-specific 3D models created from ECG-gated CT images with HeartNavigator III software. This simulation enabled the selection of a safe implant strategy and guided precise deployment, avoiding interactions between the aortic prosthesis and the Starr-Edwards mitral cage. More recently, Shu et al. [54] developed Twin-S, a DT framework specifically for skull base surgeries, which combines high-precision optical tracking and real-time simulation to integrate into image-guided interventions. Their evaluation showed that Twin-S accurately mirrors real-world drilling with an average error of 1.39 mm, enhancing situational awareness for surgeons by providing augmented surgical views. Intraoperative activity can also extend into the postoperative phase, where data and observations collected during the operation are analyzed to inform recovery strategies, assess surgical outcomes, and guide future clinical decisions, as illustrated in Fig. 4.6.

4.3.4 Medical Devices

DTs in healthcare IoT represent a paradigm shift in the management of medical devices, leveraging real-time data to enhance patient outcomes and device performance. DTs expedite continuous monitoring of devices, such as wearables and implantables, facilitating early detection of anomalies and reducing downtime using predictive maintenance. Despite these advancements, challenges such as data security, model accuracy, and interoperability persist, necessitating further research. As IoT technology continues to evolve, DTs are poised to redefine the landscape of medical device management and personalized care.

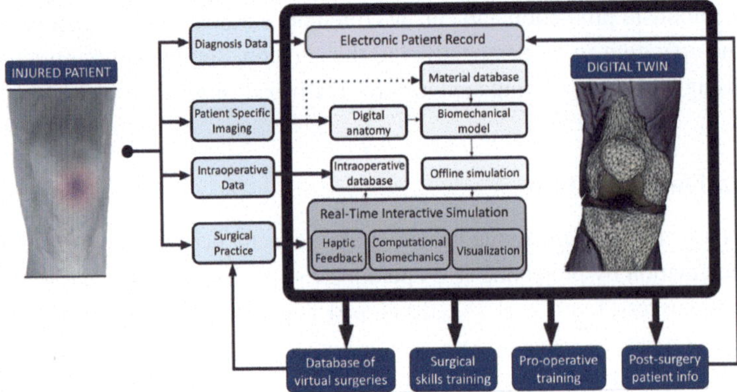

Fig. 4.6 Digital Twin Framework for Arthroscopic Knee Surgery: training, planning, and postoperative analysis [33]

Anaysis and Testing

EnvDT et al. [55] is a model-based approach designed to simulate the environmental factors of medical device DTs under uncertainty, improving the reliability of healthcare IoT application testing. It achieves approximately 61% coverage of environment models and generates diverse uncertain scenarios with a near-maximum diversity value of 0.62, ensuring more comprehensive and realistic DT-based testing. Ahmed et al. [56] highlighted how DTs can help healthcare organizations optimize medical processes, enhance patient experiences, and reduce operational costs. By continuously monitoring medical devices like X-ray and CT scan machines, DTs can mitigate system failures, ensuring reliable data collection and analysis, which is especially crucial during high-demand situations like the COVID-19 pandemic. Bersani et al. [57] explored the potential of DTs in medical systems by addressing challenges related to human behavior uncertainty and sensitive data protection. As a proof of concept, they apply their DT framework to a mechanical ventilator for COVID-19 patient care, aiming to enhance reliability, trustworthiness, and security in next-generation lung ventilators.

In the case of endoscopic medical devices, Kalozoumis et al. [58] developed a DT of the gastrointestinal tract (GIT) to provide a cost-effective and safer alternative to in vivo animal studies and clinical trials. By reconstructing intestinal geometry from medical images and creating a multiphysics model, they enable the study of device-tissue interactions, aiding in the evaluation and optimization of advanced endoscopic technologies. Another study [59] developed a DT-based simulator within a hardware-in-the-loop framework to support a clinical study on a connected medical device for lymphedema prevention in breast cancer patients after axillary lymph node dissection (ALND). By using statistical calibration and sensitivity analysis, they established measurement protocols that enabled mobile devices to detect significant signal changes, allowing early intervention and timely alerts to therapists.

Wearables and Implants

Wearable medical devices monitor health in real-time as individuals go about their daily lives, either worn on the body or attached to clothing [60]. Implantable medical devices, which replace or support biological structures, remain a key research area due to their healthcare potential [61]. Integrating these devices with DT technology enables dynamic virtual models that replicate patient conditions, optimize device performance, and enhance predictive analytics for better medical outcomes.

Chen et al. [62] used DT technology to improve healthcare monitoring in smart homes by enabling real-time visualization, anomaly prediction, and intelligent control. By combining high-fidelity DT models with wearable ECG devices and WiFi-based sensing, they develop smart algorithms for fall detection and atrial fibrillation screening, showing strong potential for more accurate and reliable healthcare monitoring. A recent work [63] proposed a DT-enabled smart system that integrates DT technology with wearable medical devices to enhance personalization in design, manufacturing, and healthcare monitoring. Through case studies on rehabilitation training, wheelchair use, and fall detection, the system demonstrates high accuracy in data tracking and customization, highlighting its potential for large-scale applications in personalized medicine. Another work [64] achieved remarkable outcomes by developing an intelligent wearable system that integrates motion and emotion recognition using multiple sensors and a novel human action recognition network (TB-SFENet), achieving 97.04% accuracy on the UCI-HAR dataset. To bridge the gap between the physical and virtual worlds, DT technology was used to create a 3D visualization platform that enables real-time interaction of activity, emotion, location, and monitoring data. Additionally, the study introduced the TGAM electroencephalogram emotion classification (TEEC) dataset with 120,000 data points, further enhancing the system's potential for applications in intelligent healthcare and virtual reality.

Yang et al. [65] proposed a decision support framework that utilizes DT technology and expert system methodologies to personalize therapy for Implantable Medical Devices (IMDs) based on real-time physiological data. By integrating a dynamic DT, reinforcement learning, and specialized algorithms, they demonstrate improved therapy customization, surpassing traditional methods in both effectiveness and patient safety, as validated through a case study on Implantable Cardioverter Defibrillators (ICDs).

Medical Metaverse

The term "metaverse" was first introduced in Neal Stephenson's 1992 science fiction novel Snow Crash. It is a blend of "meta" and "verse"—a shortened form of the universe. The integration of IoT with the medical metaverse presents innovative platforms for healthcare education and clinical training. Through the convergence of IoT-generated data streams and extended reality (XR) technologies—merging both virtual reality (VR) and augmented reality (AR), healthcare practitioners can now participate in high-fidelity clinical simulations. Ghaempanah et al. [66] developed an application that involves IoT-assisted patient simulators that provide instantaneous biometric feedback during virtual surgical procedures, helping medical trainees to develop and refine their technical skills. It also facilitates medical training, where surgeons can practice complex procedures in hyper-realistic virtual environments before performing them on real patients. This sophisticated approach to medical simulation improved the acquisition of procedural competencies and clinical readiness through experiential learning methodologies.

MeDeT [67] is a meta-learning-based approach designed to generate and adapt DTs of medical devices, addressing the challenges of integrating evolving devices in healthcare IoT testing. Evaluated in collaboration with Oslo City's health department, MeDeT achieves over 96% fidelity, adapts to new device versions in approximately one minute, and efficiently operates 1,000 DTs concurrently, providing a scalable and cost-effective alternative to physical device testing.

4.3.5 Pharmaceuticals

While the pharmaceutical industry may seem less directly connected to smart IoT technologies and DT integration, it involves several critical areas where DTs can have a significant impact. In this paper, we explore their role in drug discovery, clinical trials, and biomanufacturing, highlighting their potential to enhance efficiency, accuracy, and innovation in these domains.

Drug Discovery

DTs combined with IoT technologies, have emerged as transformative tools in drug discovery and vaccine development. Physical systems like DTs leverage real-time data to simulate biological and chemical interactions to optimize therapeutic processes in various health issues. DTs can revolutionize drug discovery by simulating the human response to different compounds, predicting efficacy and side effects before actual clinical trials. Drug discovery has traditionally relied on labor-intensive, cost-prohibitive methods. Recent advancements in digital technologies, including DTs and IoT, are changing this landscape. DTs offer dynamic and real-time simulations of biological systems, helping researchers to study drug interactions, predict outcomes, and refine therapeutic strategies. Moreover, IoT devices enhance this capability by providing continuous, real-world data streams, making DTs highly robust and adaptive. IoT devices, such as biosensors, wearable health monitors, and lab equipment, feed these simulations with real-time data, bridging the gap between in silico models and in vivo conditions. Alexandre et al. [68] illustrated the transformative potential of DTs in personalized medicine, highlighting their role in patient monitoring, risk assessment, clinical trials, and drug development. While DTs enable precise health insights and tailored treatments, they also raise ethical concerns regarding data privacy, consent, and potential biases in healthcare. Balancing innovation with ethical responsibility, they emphasize the need for robust regulations to ensure the responsible and effective integration of DTs in patient care.

Medical DTs are also transforming precision oncology by combining individual patient data with population-level insights and healthcare IoT. These models continuously monitor new data and physician decisions, allowing real-time treatment adjustments. Zhang et al. [69] portrayed that continuous feedback is particularly valuable for cancer treatment by predicting drug resistance, suggesting alternative treatments based on tumor genetics, and personalizing chemotherapy while minimizing side effects. Wu et al. [70] has shown promising results in triple-negative breast cancer treatment prediction and metastatic disease detection through radiology report analysis, which is an emerging study on cancer research. Immunology research is vital in healthcare because it helps us understand how the immune system defends against diseases, leading to breakthroughs in vaccines, autoimmune treatments, and infectious disease control. Scientists are trying to develop effective vaccines that protect against deadly pathogens like COVID-19, influenza, and tuberculosis. DTs in immunology utilize computational models and real-time data to enhance the development and manufacturing of vaccines. By integrating molecular dynamics simulations, structural biology insights, and machine learning techniques, these virtual models can predict antibody-antigen interactions, stability, and production feasibility. Recent advancements [71] have shown promising results in designing antibodies for emerging diseases and complex therapeutic targets, reducing costs and development time while enhancing therapeutic effectiveness.

Clinical Trials and Manufacturing

Clinical trial research is essential in healthcare because it ensures the safety, efficacy, and effectiveness of new medical treatments, drugs, and medical devices before they reach patients [72]. These trials provide scientific evidence to determine whether a new intervention works better than existing treatments or has fewer side effects. One of the most promising applications of DTs in clinical trials is the development of virtual control groups, which reduce the need for placebo arms while maintaining statistical robustness [73]. Sel et al. [74] proposed an approach that enhances patient recruitment and retention while speeding up trial completion. Additionally, this research supports adaptive trial designs, where real-time patient data dynamically modifies study parameters. Recently, regulatory agencies, such as the FDA and EMA, have explored DT-based trial validation frameworks in healthcare that will be expected to redefine precision medicine and accelerate the development of life-saving treatments. TWIN-GPT [75] is a large language model-based approach for creating personalized DTs that establish cross-dataset associations of medical information, enabling more accurate clinical trial outcome predictions even with limited data. Through comprehensive experiments, we demonstrate that TWIN-GPT outperforms previous methods in prediction accuracy and generates high-fidelity trial data, showcasing its potential to enhance virtual clinical trials and advance healthcare research.

In the specialized field of biomanufacturing, DT innovation is transforming multiple aspects of production, including process development, validation protocols, and operational management. The application of DTs in biomanufacturing creates an intricate, mathematically driven model that captures every aspect of the production process. This comprehensive simulation spans the entire manufacturing journey, beginning with raw material inputs and continuing through to the final product output. Herwig et al. [76] highlighted that DTs have expanded beyond biomolecule production into chemical synthesis. This system's sophistication lies in its ability to continuously integrate both historical process data and real-time information, monitoring and analyzing crucial parameters such as temperature fluctuations, pH levels, and nutrient concentration variations for drugs.

4.3.6 IoMT Security and Privacy

The rapid adoption of the IoMT has significantly improved healthcare delivery by enabling real-time monitoring, remote diagnostics, and personalized treatments [77]. However, the increasing interconnectivity of medical devices introduces substantial cybersecurity risks, including data breaches, unauthorized access, and system vulnerabilities that could compromise patient safety. For instance, a security breach in data collected by a smartwatch or fitness tracker could expose sensitive personal details such as a user's health metrics, daily activities, and demographic information [78]. Securing IoMT ecosystems requires robust measures across data collection, transmission, and storage, especially as DT technology becomes more integrated into healthcare, further emphasizing the need for resilient cybersecurity frameworks. In an IoMT environment, DT solutions demand robust security across all layers—sensor, gateway, cloud, and visualization—to safeguard patient data and ensure system integrity, as depicted in Fig. 4.7. The figure illustrates how IoMT sensors interface with edge computing nodes and secure protocols, forming a critical foundation for real-time data exchange and effective healthcare monitoring.

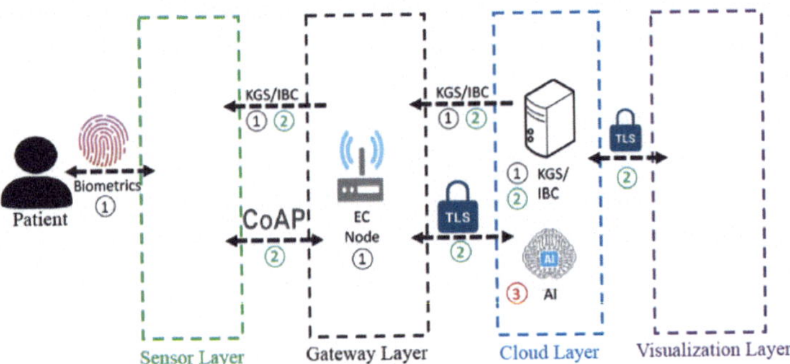

Fig. 4.7 A proposed IoMT secure system architecture [79]

Device and Network Security

Zachos et al. [80] presented an innovative hybrid anomaly-based Intrusion Detection System (AIDS) designed to address security vulnerabilities in IoMT networks. Additionally, they developed a functional IoT/IoMT security testbed, enabling the assessment of AIDS against various cyber threats and providing a valuable tool for researchers working on IoT security solutions. DT-IoMT [81] is a DT reference model designed to integrate the human body and medical devices within IoMT, enabling real-time health monitoring and secure, autonomous diagnostics. This paper presents a conceptual architecture that incorporates smart sensors, advanced communication protocols, and biomedical signal monitoring while addressing security challenges, making it a foundation for future implementations in chronic disease management and smart hospitals.

In the case of cyber resilience, a study [82] proposed a new deep learning scheme to automatically detect vulnerabilities in healthcare DTs by capturing bi-directional context relationships among risky code keywords. Through extensive experiments on IoT-related software, the approach outperforms state-of-the-art methods, addressing the complexity and scale of emerging 6G healthcare DT applications. Jimenez et al. [83] explored the implementation of Wireless Body Area Networks (WBAN) in healthcare, highlighting performance and security challenges, especially concerning wearable or implanted devices. They also provide definitions of Medical Cyber-Physical Systems (MCPSs) and DTs, discussing how these technologies, alongside IoT and cloud computing, can improve patient care. Sharma et al. [84] proposed a novel approach that merges elliptic curve cryptography (ECC) with blockchain technology to secure IoT-based healthcare DTs, while employing a genetic algorithm-optimized random forest (GAO-RF) to enhance feature selection. They demonstrated robust performance, achieving an F1-score of 97.3% and an accuracy of 98.4%, thereby instilling trust in the healthcare system by effectively safeguarding patient data.

Data Privacy and Regulatory Compliance

Ensuring robust data protection and regulatory adherence is critical for maintaining trust in digital healthcare systems. By incorporating strong privacy measures and compliance frameworks, healthcare providers can safeguard patient information while supporting innovative solutions like DTs. Lutze et al. [85] presented a novel approach for managing AI-driven eHealth systems using DTs, categorized into personal, group, and system-level twins. This framework enables compliance with EU-GDPR by allowing the unlearning of personal data, enhances system transparency to ensure unbiased conclusions, and incorporates self-monitoring mechanisms to reassess prior diagnoses based on newly acquired knowledge. Akash et al. [86] proposed a mathematical data model to systematically collect and organize patient information, ensuring privacy and security. Building on this model, they introduced a healthcare DT system underpinned by blockchain, explaining its components and protocol flows. Finally, they evaluated its feasibility by comparing it with other relevant approaches. PSim-DTH [87] is an efficient and privacy-preserving scheme that employs a partition-based tree (PB-tree) and matrix encryption to securely handle healthcare data on a DT cloud platform. By enabling similarity queries without revealing personal information, it supports continuous patient health monitoring while safeguarding sensitive details. The scheme's evaluation demonstrates its strong data protection capabilities and high query efficiency.

Another work [88] proposes a DT-enabled Asynchronous SplitFed Learning (DT-ASFL) framework for e-healthcare classification, utilizing model splitting and feature-only data transmission to handle heterogeneous IIoT resources and privacy constraints. By employing DT for real-time device monitoring and asynchronous updates, the approach enhances communication efficiency and accommodates resource-constrained devices, thus improving performance over existing methods. Tang et al. [89] explored a blockchain-based framework to enhance the security and privacy of healthcare DT data (HDTD) sharing, addressing concerns related to data accessibility and integrity. By implementing an access control scheme with cloud storage and attribute encryption, they enable secure interactions between users. Additionally, they develop an HDTD missing value prediction algorithm to mitigate data loss and ensure real-time data reliability, demonstrating improved security and reduced interaction delays compared to existing methods.

4.3.7 Other Miscellaneous Applications

In addition to the literature reviews presented in the previous sections, several other significant studies deserve attention but were not explicitly categorized within those sections. These works focus on key areas such as comprehensive system architecture, advancements in e-health, and innovative approaches to healthcare monitoring for smart home systems.

Comprehensive Architecture

Several studies have proposed comprehensive architectures for health-related DTs across various domains, aiming to address multiple challenges concurrently. Some examples [1, 52, 81] have already been highlighted in the personalized sections. Benedictis et al. [90] reviewed the current literature on DT in healthcare, categorized its applications into four main domains, and proposed a reference DT architecture. They then introduce CanTwin—a real-life canteen service DT developed by Hitachi—for social distancing, queue inspection, and occupancy monitoring, demonstrating the practical utility of DT in virus containment. Iqbal et al. [91] explored the potential of DTs in healthcare by addressing key challenges related to their design, components, and operational requirements. By proposing an implementation architecture and presenting a proof-of-concept prototype, they demonstrate how DTs can enhance healthcare services while highlighting open challenges and future research directions.

Pellegrino et al. [92] conducted a meta-review using the PRISMA methodology to analyze 20 selected reviews from 1,075 studies, identifying key functional, technological, and operational aspects of DTs in healthcare. Based on this analysis, they develop a conceptual framework that clarifies the hierarchical relationships among these aspects, facilitating the effective design, evaluation, and implementation of DTs in digital health. Alqahtani et al. [93] proposed a secure framework that integrates DT technology, IoT-edge computing, and blockchain to monitor adults' physical activity while ensuring the privacy of health data. By leveraging deep learning-assisted CNN and LSTM models for real-time vulnerability assessment, their approach outperforms state-of-the-art techniques in prediction accuracy, reliability, and stability.

E-health

Since the DT concept has now expanded to include both living and inanimate entities, its potential for use in eHealth and wellness applications has significantly grown [94, 95]. CloudDTH [96] is a novel cloud-based healthcare framework designed for real-time monitoring, diagnosing, and predicting health conditions using wearable medical devices. It bridges physical and virtual healthcare spaces through a DT model, offering enhanced personal health management and real-time supervision, especially for elderly patients. Garg et al. [97] proposed a system integrating DTs and CPS to gather and analyze real-time patient data via wearable health instruments, maintaining parallel physical and virtual databases for diagnosis and monitoring. They further secure e-health cloud data using a two-phase iris biometric authentication method based on an EfficientNet CNN, effectively differentiating genuine from spoofed samples. Another work [98] expanded on the DT concept for health and medical software by incorporating regulatory standards, unique device identification (UDI) systems, and full lifecycle management across personal, group, and system-level twins. They also explore the strengths and weaknesses of using DTs in software design and development, emphasizing the benefits of agile approaches over traditional V-model methods. Furthermore, DT-ASFL [88], which we discussed before, improves learning efficiency in IIoT-based healthcare by enabling asynchronous model updates and partial model training, allowing resource-constrained devices to participate while maintaining high performance compared to existing methods.

Healthcare Monitoring for Smart Home

DT is increasingly being explored for healthcare monitoring in smart homes, enabling real-time visualization, anomaly detection, and intelligent control. By integrating DT with wearable sensors and AI-driven analytics, researchers aim to enhance remote patient monitoring, predict health risks, and improve personalized care, ultimately fostering a smarter and more responsive home healthcare environment [99]. Akram et al. [100] introduced the AI-Generated Optimal Decision (AIGOD) algorithm and the Deep Diffusion Soft Actor-Critic (DDSAC) framework to enhance the integration of HDTs with AI-Generated Content (AIGC) in IoMT-based smart homes. Their AI-generated content-as-a-service (AIGCaaS) architecture, optimized for IoMT environments, demonstrated superior performance, with DDSAC achieving a 20% improvement in task completion rates and a 15% increase in overall utility, highlighting its potential for personalized and proactive healthcare interventions. Another interesting work [101] proposed a DT-based model for home-device (HD) control, integrating IoT, virtual simulation, and remote intelligent control to establish a human cyber-physical system (HCPS). By utilizing a game theory approach for modeling IoT-based smart devices and enhancing sensing networks, they improve virtual reality synchronization, positional accuracy, and quality control in HDDT development.

Rivera et al. [102] investigated how DT may transform healthcare by facilitating ongoing monitoring and enhancing communication between patients, caregivers, and systems. Through virtual health tracking and therapy evaluation, they hope to improve precision medicine by combining DT with data-driven techniques like machine learning. Their research presents a reference model that uses DT, autonomic computing, and self-adaptive systems to create intelligent, adaptable healthcare solutions that facilitate treatment customization and decision-making. Another work [103] proposed a context-aware framework that uses DT to monitor indoor air quality and human activity, integrating IoT, 6G networks, edge computing, and AI-driven analytics. By developing an architecture with sensors, gateways, and a DT object on the Azure cloud, they enhance healthcare monitoring through real-time data processing and predictive insights. Their contributions include linking 6G sensing with IoT techniques and achieving high-accuracy human activity recognition using deep learning models. Furthermore, there are studies [87, 96] that used the cloud-based healthcare monitoring system discussed previously.

4.3.8 Summary of DT Applications in Healthcare IoT

To synthesize the findings across the diverse application domains discussed in Sects. 4.3.1 to 4.3.7, one can observe that DT applications in healthcare IoT are multifaceted. Some works focus on smart hospital management, using twins to improve operational efficiency or environmental safety. Others center on patient-specific modeling for personalized care (such as the diabetes and cardiac twins), leveraging real patient data streams to tailor healthcare decisions. Yet others target medical devices and infrastructure, ensuring that the underlying IoT-enabled systems (from wearable sensors to large imaging devices) are reliable, secure, and optimized. Evaluation metrics across studies vary according to these goals—ranging from technical performance indicators (latency, detection accuracy, simulation error) to clinical/proxy outcomes (like predicted health metrics or workflow improvements). Each approach brings certain strengths: for instance, data-driven twins with machine learning excel at real-time prediction, while physics-based twins offer interpretability and physiological insight. However, limitations are also evident. Common challenges include limited clinical validation (many are prototypical demonstrations or simulations), scalability issues (handling the volume and velocity of IoT data in real hospital settings), and interoperability and privacy concerns (integrating diverse IoT devices and ensuring patient data security). These observations set the stage for a deeper discussion on open research questions and the future directions needed to advance DT technology in healthcare.

4.4 Healthcare DT Enabling Technologies

4.4.1 Modeling and Simulation

A simulation model remains static unless the designer manually adds new components. In contrast, a DT initially appears similar to a simulation model but acquires dynamic properties by integrating real-time data, enabling it to adapt and evolve over time [104]. This real-time data feeds into sophisticated AI-driven predictive models, enabling disease trajectory forecasting and treatment optimization. Elayan et al. [1] implemented a model that successfully predicted a particular heart condition with high accuracy in different algorithms. The findings showed that integrating DT with the healthcare IoT field would improve healthcare technologies by bringing patients and healthcare professionals together in a robust, cost-effective, and scalable health ecosystem.

Data-driven approaches, i.e., machine learning algorithms, deep learning models, and advanced AI techniques—including Generative AI (Gen AI), artificial intelligence-generated content (AIGC), and Large Language Models (LLMs)—are increasingly being leveraged to prototype and develop high-impact DTs across various sectors within healthcare. As Gen AI continues to advance and shape the future, Fig. 4.8 provides a high-level overview of its integration with DTs in healthcare. Applications of various approaches include but are not limited to—validation using gradient boosting regressor [105], LLM-based DT [75], GAN based DT [51], long-term memory (LSTM) model for predictive analysis [1, 106],

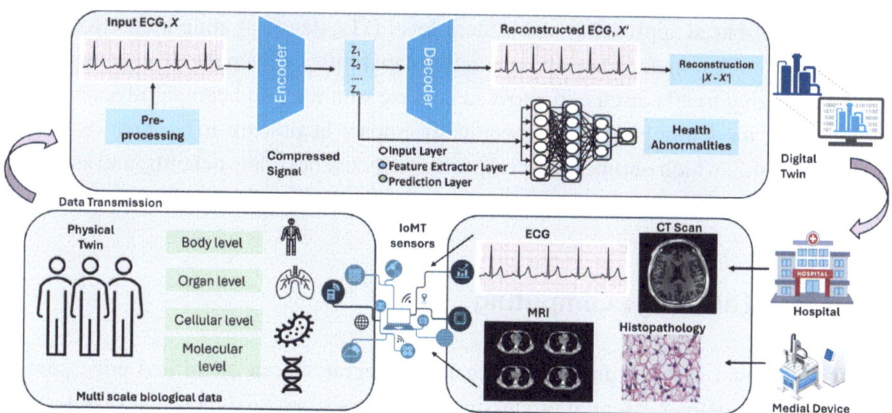

Fig. 4.8 An overview of Gen AI-based digital twins in healthcare with a focus on ECG data as an illustrative example

Table 4.2 Comparison of modeling and simulation paradigms for DTs in healthcare IoT

Approach	Contributions	Effectiveness	Limitations
Data-driven Models	Real-time predictive analytics, personalized patient monitoring, anomaly detection	Highly effective in data-rich environments; excels at personalized health monitoring and rapid anomaly detection	Requires substantial amounts of high-quality data; privacy and ethical challenges; limited interpretability ("*black-box*")
Physics-based Models	Provides mechanistic insights, scenario simulation, supports "*what-if*" analysis	Effective in scenarios requiring physiological fidelity and causality (e.g., surgical planning, device testing)	Computationally intensive; demands extensive patient-specific data for model calibration; challenging real-time deployment
Hybrid Models	Combines data-driven accuracy and physics-based interpretability; suitable for complex, multi-modal systems	Highly effective in balancing predictive power and interpretability; adaptable and scalable for complex healthcare scenarios	Complex implementation; integration challenges; difficult model validation and debugging due to multiple interacting components

improved Random Forest for clinical big data analytics [107], support vector machine (SVM) for stroke diagnosis [108], customized CNN [82], mobile AIGC-driven HDT [109], Gen AI with 3D modeling [110] etc. Furthermore, several impactful studies [111, 112] have explored model-based approaches and system-level DTs, demonstrating their effectiveness in enhancing simulation accuracy and predictive capabilities. Given the diverse methodologies and emerging trends discussed above, a concise summary and comparative analysis of how these DT modeling paradigms specifically impact healthcare IoT scenarios are provided in Table 4.2, which outlines their respective effectiveness, key benefits, and associated challenges, facilitating a clearer understanding of their applicability.

4.4.2 Cloud and Edge Computing

The computational architecture supporting DTs integrates both cloud and edge computing paradigms to deliver essential processing power, real-time analytics, and seamless data synchronization. Cloud computing plays a crucial role by offering vast storage capacity, facilitating AI model training, and enabling large-scale data analysis, which is particularly valuable for long-term patient monitoring, electronic health records (EHR) management, and predictive diagnostics in hospitals. Meanwhile, edge computing ensures immediate data processing at the source, reducing latency and enabling real-time decision-making for critical applications such as intensive care unit (ICU) monitoring, remote patient monitoring, and computer-assisted surgical procedures [113]. By using IoT-enabled edge computing, hospitals can process patient data locally, reducing dependence on cloud networks and ensuring rapid response times, which is essential for emergency care and high-risk patient management. This synergy between cloud and edge computing enhances the efficiency of DT-based healthcare systems by optimizing resource allocation, improving data security, and ensuring compliance with regulatory standards such as HIPAA and GDPR. Alsuba et al. [106] proposed an intelligent athlete healthcare framework using IoT-edge computing to monitor real-time health data and assess probabilistic health state susceptibility during intense training. By integrating a Bayesian classification model and a multi-scale LSTM model for predictive analysis, they demonstrate superior performance over existing techniques in terms of classification efficiency, predictive accuracy, and system stability.

4.4.3 Blockchain Technology

To ensure service sustainability, a trustworthy and shareable ledger, commonly known as the blockchain, is maintained for certain DT systems [114]. Blockchain compiles dispersed, secure, and verifiable records of data collected from multiple sources, linking them in a single chain of interconnected blocks [115, 116]. It plays a vital role in DT healthcare frameworks by enabling secure, immutable, and distributed data exchange while addressing key requirements such as data security, privacy protection, and regulatory compliance [117]. Some applications are focused on data privacy, as we discussed in Sect. 4.3.6. Key studies have explored the use of blockchain for various applications, including privacy and security [86, 89], integration with ECC [84], irregular event analysis [118], and public auditing schemes [119], among others.

4.5 Discussion

According to this review, DT research in healthcare systems is experiencing rapid global growth, focusing on three primary objectives: current trends, healthcare innovation, and applications. However, the field faces two key challenges: an unclear definition of essential DT characteristics of data management and a limited exploration of IoT modeling techniques for healthcare applications. Recent trends indicate that most DT studies focus on problem formulation and conceptualization, with a substantial portion dedicated to enhancing patient care without considering real-world applications. Despite this initial approach being both fundamental and relevant, we argue that expanding DT applications of healthcare systems and understanding provider decision-making processes could significantly enhance care quality and system resilience.

4.5.1 Current Trends

Since it was proposed in 2002 by Professor Grieves, DT has expanded its healthcare research [120], covering various sources, indicating multidisciplinary interest and applications. We have analyzed the recent trends in significant research publications that focus on DTs in healthcare IoT over the past few years. Figure 4.9 The pie chart presents the distribution of research papers on DT research in healthcare from 2012 to 2024 across various academic journals and conference proceedings. The largest share of publications appears in IEEE Access, which accounts for 17.4% of the total research output in this field, followed closely by Sensors at 14.9%. Several Lecture Notes series, including those in electrical engineering, networks and systems, computer science, and civil engineering, collectively represent a substantial portion of the published work, indicating their significance in disseminating DT research. Additionally, other important publication venues include Scientific Reports, Applied Sciences, Journal of Physics: Conference Series, and Mechanical Systems and Signal Processing, each contributing between 6 and 8% of the total papers. The distribution of these publications highlights the interdisciplinary nature of DT research in healthcare, spanning fields such as engineering, computer science, and applied sciences in healthcare.

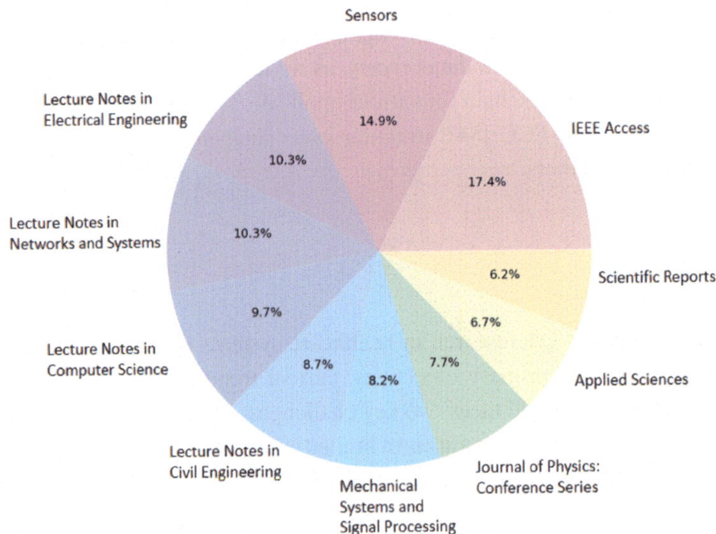

Fig. 4.9 Major domains in digital twin research in healthcare from 2012 to 2024

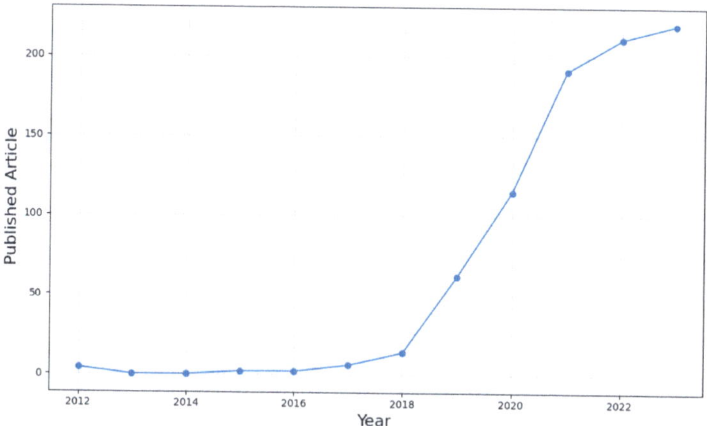

Fig. 4.10 Digital twin research trends in healthcare from 2012 to 2024

This visualization provides valuable insights for researchers looking to identify relevant journals and conferences for publishing their work, as well as for tracking trends in the field over time.

We also conducted a year-by-year search for articles using the keyword "digital twin in healthcare" while including health-related studies. This approach aimed to capture the overall growth of DT technology in the literature, in terms of the healthcare domain. Figure 4.10 illustrates the number of documents published over time from 2012 to 2023, showing a significant upward trend in recent years. From 2012 to around 2017, the number of documents remained consistently low, with minimal fluctuations. However, starting in 2018, there was a gradual increase, followed by a sharp exponential rise from 2019 onward. The publication count peaked around 2022-2023, indicating a surge in research activity. This pattern suggests growing interest and advancements in the subject area, with an accelerating rate of publications in recent years.

4.5.2 Emerging Trends

Recent developments reveal novel applications and underexplored domains for DTs in healthcare IoT that extend beyond the scope of prior surveys. One notable domain is mental health, where DT technology remains largely untapped. The use of DTs in mental health is "*yet to be explored*", indicating a clear research gap. While most healthcare DT efforts focus on patient-level or device-level twins, DTs can also model entire clinical workflows and healthcare facilities. For example, a DT of a hospital or clinic can optimize patient flow, predict ICU bed occupancy, and manage resource allocation in real-time. A recent review noted that using DTs to manage healthcare systems is an "*emerging topic*" supported by limited literature so far [30]. Nevertheless, early implementations demonstrate potential:

hospital-level twins are being used to simulate emergency department workflows and identify bottlenecks, with significant promise for improving operational efficiency. The COVID-19 pandemic catalyzed interest in such system-level twins, as they could help model and respond to surges in patient volume and resource needs. This trend underscores a broader vision of population-level or public health DTs, where entire patient populations or health systems are replicated virtually to test policies and pandemic responses in a risk-free environment.

As discussed earlier in Sect. 4.4.1, the integration of advanced AI techniques—particularly generative AI (e.g., GANs, LLMs, AIGC) and federated learning—has already begun to transform the development of DTs in healthcare. These technologies enable high-fidelity virtual modeling, privacy-preserving distributed training, and synthetic data generation, especially valuable in data-scarce or privacy-sensitive clinical environments. A recent survey [121] emphasizes that "*generative AI may be a promising solution*" for improving digital human twins by augmenting digital models with realistic data while accurately reflecting real-time physical states. Looking forward, sustained progress in GenAI and federated architectures will be crucial for scaling adoption, enhancing model generalizability, and maintaining ethical compliance. Future research should prioritize validation, transparency, and regulatory alignment for these AI-enhanced DTs.

4.5.3 Architecture and Infrastructure Implications

As outlined in Sect. 4.4.2, a *layered edge-cloud architecture* is now the de facto design pattern for healthcare DTs. In short, time-critical analytics (e.g., real-time vital-sign monitoring) run on local edge servers to minimize latency and keep sensitive data on-site, while computationally heavy tasks—such as high-fidelity organ simulation or cross-patient model retraining—are offloaded to secure cloud resources. Discussion here, therefore, focuses on the deployment implications that arise from this pattern.

- **Interoperability**: Successful scaling hinges on adopting open healthcare data standards and middleware that can translate heterogeneous IoMT data streams into those formats. Standards-compliant middleware future-proofs the platform: new sensors or EMR systems can be integrated without re-engineering the twin.
- **Security and Privacy**: Because twins aggregate high-value clinical data, end-to-end protection is mandatory. Recommended safeguards include transport-layer encryption, role-based access control, and tamper-evident ledgers (e.g., blockchain) for model-update provenance [80]. For multi-institutional learning, *federated analytics* or secure multi-party computation can keep raw data local while still improving a global twin.
- **Network Readiness**: Emerging 5G/6G links offer the ultra-low latency and bandwidth required for continuous, bidirectional streaming between devices, edge nodes, and cloud datacentres-an essential enabler for closed-loop twins.

- **Take-away**: A practical healthcare DT stack must therefore combine (i) edge servers for on-premises real-time analytics, (ii) standards-based middleware for data normalization, and (iii) security-by-design at every layer. These architectural pillars transform the abstract concept of DTs into a deployable, scalable clinical technology.

4.5.4 Healthcare Data Standards

The successful implementation of DTs in healthcare IoT hinges on the adoption of established data standards—namely HL7, FHIR, and the OMOP Common Data Model—to ensure seamless data exchange and consistent representation of patient information [122]. These standards enable a secure, interoperable environment where real-time updates from multiple healthcare systems can be compiled into accurate digital replicas. Dolin et al. [123] introduced HL7 as an international set of standards designed to streamline healthcare information exchange, creating a structured framework for efficiently sharing, integrating, and retrieving electronic health data. Building on HL7 principles, the Fast Healthcare Interoperability Resources (FHIR) standard modernizes data transactions by incorporating web technologies, thereby facilitating the continuous flow of real-time clinical information into digital models. This capability is vital for applications that rely on current patient data from diverse providers and systems.

Equally important, Makadia et al. [124] developed the OMOP Common Data Model, offering a unified structure for storing and querying healthcare data. By converting heterogeneous data sources into a consistent format, DT platforms can more effectively process and analyze large-scale patient information, ultimately yielding deeper insights into health patterns and outcomes. In this way, HL7, FHIR, and OMOP CDM collectively form the foundational layer that enables DTs in healthcare IoT—ensuring data integrity, interoperability, and real-time responsiveness.

4.6 Future Directions and Challenges

Although this discussion does not serve as a comprehensive methodological review of DT technologies, it is important to acknowledge and highlight several well-documented challenges that continue to hinder advancements in this field. Many implementations remain pilot projects or lab demonstrations, and there is a significant journey ahead to achieve widespread clinical adoption. In fact, a recent review found that about 98% of purported healthcare DTs are still in preclinical phases, underscoring that most work has not yet translated into routine clinical practice [125]. In this section, we outline the key open research questions and persistent technical and clinical challenges that must be addressed. We also provide recommendations for researchers and practitioners aimed at propelling this field forward.

4.6.1 Open Research Questions and Gaps

How to Achieve High-Fidelity, Validated Digital Twins?
One fundamental open question is how to ensure that a DT accurately reflects the complex reality of a patient or system over time. Rigorous validation against clinical outcomes is often lacking. Researchers are asking what level of modeling detail is truly necessary for a twin to be *trustworthy* in decision-making, and how to validate these models prospectively in clinical trials or real-world deployments.

What are the Generalizable Frameworks or Benchmarks?
As the field grows, there is a need to move from bespoke solutions to general frameworks. An open question is whether a *common reference architecture* (or set of design patterns) for healthcare IoT twins can be defined, to avoid each project reinventing the wheel. Similarly, the lack of benchmarks and shared datasets makes it hard to compare approaches.

How to Enable Two-Way Interaction and Control?
Many current DT implementations are largely one-way (streaming data from the patient/device to the twin). Only a minority (around 11% in one analysis) support full *bi-directional* data flow or closed-loop control [125]. It is still an open research problem to develop twins that not only passively reflect the physical system but also actively intervene (e.g., sending control signals to IoT devices or real-time decisions in a feedback loop).

What are the Long-Term Impacts and Emergent Behaviors?
As DT systems become more complex (incorporating AI, simulating physiology, etc.), unforeseen behaviors might emerge. Researchers are curious about the long-term performance: *Will a twin remain accurate as a patient's condition evolves? How do we update models to avoid drift?* Also, if multiple DTs interact, what are the system-level effects? [126].

Addressing these open questions will likely require interdisciplinary efforts, combining insights from computer science, biomedical engineering, and even fields like complex systems theory and human factors. Each question points to a gap in knowledge that future research must fill to move DTs from promising prototypes to reliable tools in healthcare IoT.

4.6.2 Challenges

From an engineering standpoint, implementing DTs in the healthcare IoT context presents several technical challenges:

Data Management and Quality
Healthcare IoT devices generate enormous amounts of data, often continuously. Ensuring data quality, consistency, and completeness is non-trivial-sensors can fail or drift, and biomedical data can be noisy. Techniques for real-time data filtering, anomaly detection, and handling missing data are critical. Scalable database solutions and edge computing for preprocessing are active areas of development to address this challenge.

Interoperability and Standardization
The current ecosystem of medical IoT devices is highly fragmented. Devices from different manufacturers use different data formats and communication protocols. A DT that aggregates data from, for example, a wearable heart monitor, an implantable device, and a smart hospital bed must interface with all seamlessly. The lack of standard data models and APIs in healthcare IoT hampers integration. Existing standardization efforts need to be extended to support the real-time streaming data required by DTs.

Model Accuracy Versus Complexity Trade-Off
There is a fundamental tension between DT fidelity and computational complexity. High-fidelity physics-based models can be very accurate but computationally expensive, whereas simpler data-driven models are faster but may miss important physiological or system-level nuances. Multi-level modeling approaches-combining fast coarse models with detailed models invoked on demand-offer a compromise but introduce additional orchestration complexity.

Cybersecurity
DTs expand the attack surface by interfacing with numerous IoT endpoints and aggregating highly sensitive healthcare data. Protecting the integrity of the twin is critical to prevent data poisoning, false state injection, or malicious manipulation of clinical decisions. While specialized security frameworks for medical IoT and DT environments are emerging, achieving robust end-to-end security remains an open technical challenge.

On the clinical side, there are significant challenges in translating DT innovations into practice:

Clinical Validation and Evidence Generation
A major barrier to adoption is the lack of rigorous clinical validation for most DT solutions to date. Healthcare is inherently evidence-driven, requiring well-designed clinical studies to demonstrate that DT-based systems improve patient outcomes, safety, or operational efficiency. In the absence of strong empirical evidence, clinicians may remain hesitant to trust or adopt these technologies.

Integration into Clinical Workflow
Even when a DT tool demonstrates technical effectiveness, integrating it seamlessly into existing clinical workflows remains challenging. Key issues include designing clinician-friendly user interfaces, determining how and when DT outputs should be presented to care teams, and minimizing false alarms or clinically irrelevant recommendations. Striking the right balance between decision support and automation is also critical; for example, whether a DT should autonomously adjust a ventilator setting or merely recommend actions to a physician.

Interdisciplinary Communication
The successful development and deployment of DTs depend on close collaboration between technical experts (e.g., engineers and data scientists) and clinical practitioners (e.g., physicians and nurses). Differences in terminology, priorities, and professional culture can impede mutual understanding and trust. Building interdisciplinary teams and cultivating roles such as *clinical data scientists* can help bridge this gap and improve translational impact.

In summary, while the technical development is critical, clinical adoption will depend on trust, evidence, and ease of use. Overcoming these challenges requires not just technological innovation, but also workflow design, medical research, and clear demonstrations of value in patient care.

4.6.3 Recommendations for Advancing the Field

To overcome the above challenges and knowledge gaps, we propose several future directions and recommendations for researchers, clinicians, and other stakeholders working on DTs in healthcare IoT:

Focus on Interdisciplinary Collaboration

The development of effective healthcare DTs should involve a tight collaboration loop between technologists and healthcare practitioners. We recommend forming interdisciplinary teams that include software/AI engineers, clinicians, biomedical scientists, security experts, and even patients. Such teams can ensure that the solutions address real clinical needs, are technically sound, and ethically grounded. Regular workshops or joint "*clinician-in-the-loop*" design sessions can help align the twins' functionality with frontline medical workflows.

Develop open Platforms and Standards

To accelerate progress, the community would benefit from open reference architectures and shared tools. We encourage the creation of open-source platforms for DT development specific to healthcare, where common services are provided. This would let researchers focus on novel modeling aspects without rebuilding infrastructure. In parallel, engaging with standards organizations to push for interoperability standards will reduce integration friction. A concerted effort toward standardizing how IoT data feeds into DT models and how results are output will make it easier to adopt these systems across different hospitals and device ecosystems.

Strengthen Validation and Publish Negative Results

Academic and industry researchers should prioritize rigorous validation studies and openly share results. This includes not only publishing success stories but also *negative or inconclusive findings*, which are equally informative. For instance, if a certain approach to modeling failed to improve outcomes, reporting it can save others time and guide the field away from blind alleys. We recommend establishing *benchmark datasets and challenge competitions* (as in AI fields) for tasks like predicting patient deterioration can drive progress. Collaborating with clinical units on pilot studies will help build the needed evidence base and accelerate learning from real-world results.

Invest in Scalability, Security, and Robustness Research

As identified, technical robustness is a major challenge. Funding and research efforts should be devoted not only to novel applications but also to the *"boring"* but crucial aspects of scalability and security. This means developing algorithms that can handle big data with constrained resources (perhaps using edge computing for initial data crunching) and exploring advanced security measures. For example, incorporating blockchain or distributed ledger techniques for tamper-proof logging of twin data transactions or using differential privacy methods to allow data analysis with mathematical privacy guarantees. We also recommend stress-testing DT systems under worst-case scenarios (network down, data corrupted, etc.) to ensure failsafes are in place. These engineering investments might not be as glamorous as new AI models, but they are vital for deployment in the safety-critical healthcare arena.

Engaging all Stakeholders and Educating End-Users

Finally, we recommend that hospital administrators and policymakers be informed of DTs' potential to improve outcomes and reduce costs, using case studies or ROI simulations to gain support. Clinicians and staff need training to understand and interpret DT outputs through programs or CME modules. Patient education is equally important—introducing the concept via brochures or consent discussions can help build trust and avoid confusion when DTs are used in care. Informed stakeholders and users enable more effective adoption and impact.

4.7 Conclusion

DT technology is rapidly transforming the healthcare landscape, offering unprecedented opportunities for personalized medicine, efficient hospital management, and enhanced patient care. By integrating real-time data with IoT infrastructure, DTs enable dynamic patient modeling, predictive diagnostics, and optimized medical device performance. Additionally, their role in the pharmaceutical industry, from clinical trial simulations to supply chain monitoring, highlights their broader impact on healthcare innovation. Amid these technological advancements, this paper has provided a systematic review of DT integration within the healthcare IoT ecosystem, emphasizing its critical role in enabling personalized medicine and addressing pressing security challenges. We have explored the applications of DTs in digital patient modeling, smart hospital lifecycle management, medical device optimization, and pharmaceutical advancements, demonstrating their transformative potential in modern healthcare.

Despite these advancements, several challenges remain, including data security, interoperability, and the ethical implications of patient data usage. Addressing these concerns is crucial for ensuring the widespread adoption and reliability of DTs in healthcare. Moving forward, continued research and development in AI-driven analytics, secure data-sharing mechanisms, and regulatory frameworks will be essential to unlocking the full potential of DTs in smart healthcare systems. By overcoming these barriers, DTs can pave the way for a more efficient, proactive, and patient-centered healthcare ecosystem.

References

1. Elayan, Haya, Moayad Aloqaily, and Mohsen Guizani. 2021. Digital twin for intelligent context-aware IoT healthcare systems. *IEEE Internet of Things Journal* 8 (23) :16749–16757.
2. Hussain, Adedoyin Ahmed et al. 2020. Notice of retraction: AI techniques for COVID-19. *IEEE Access* 8: 128776–128795.
3. Kumar, Pardeep et al. 2024. *Healthcare industry assessment: Analyzing risks, security, and reliability*.
4. Fuller, Aidan, et al. 2020. Digital twin: Enabling technologies, challenges and open research. *IEEE Access* 8: 108952–108971.
5. Groth, Corrado et al. 2018. The medical digital twin assisted by reduced order models and mesh morphing. In *International CAE conference*, vol. 10.
6. Chu, Yanting Chu et al. 2023. The potential of the medical digital twin in diabetes management: A review. *Frontiers in Medicine* 10: 1178912.
7. Coorey, Genevieve et al. 2021. The health digital twin: Advancing precision cardiovascular medicine. *Nature Reviews Cardiology* 18 (12): 803–804.
8. Coorey, Genevieve et al. 2022. The health digital twin to tackle cardiovascular disease—a review of an emerging interdisciplinary field. *NPJ Digital Medicine* 5 (1): 126.
9. Vats, Tarun et al. 2023. Explainable context-aware IoT framework using human digital twin for healthcare. *Multimedia Tools and Applications* 1–25.
10. Okegbile, Samuel D., et al. 2022. Human digital twin for personalized healthcare: Vision, architecture and future directions. *IEEE Network* 37 (2): 262–269.

11. Barricelli, Barbara Rita et al. 2020. Human digital twin for fitness management. *IEEE Access* 8: 26637–26664.
12. Graessler, Iris and Alexander Pöhler. 2017. Integration of a digital twin as human representation in a scheduling procedure of a cyber-physical production system. In *2017 IEEE international conference on industrial engineering and engineering management (IEEM)*, 289–293. IEEE.
13. Löcklin, Andreas et al. 2021. Architecture of a human-digital twin as common interface for operator 4.0 applications. *Procedia CIRP* 104: 458–463.
14. Wang, Baicun et al. 2022. Human digital twin (HDT) driven human-cyber-physical systems: Key technologies and applications. *Chinese Journal of Mechanical Engineering* 35 (1): 11.
15. Wang, Baicun et al. 2024. Human digital twin in the context of industry 5.0. *Robotics and Computer-Integrated Manufacturing* 85: 102626.
16. *Healthcare Digital Twins Market Size—Industry Report, 2030*. 2024. Accessed 01 Jan 2025. https://www.grandviewresearch.com/industry-analysis/healthcare-digital-twins-market-report.
17. Vishnu, S., SR Jino Ramson, and R. Jegan. 2020. Internet of medical things (IoMT)-An overview. In *2020 5th international conference on devices, circuits and systems (ICDCS)*, vol. 2020, 101–104. IEEE.
18. Catarinucci, Luca et al. 2015. An IoT-Aware architecture for smart healthcare systems. *IEEE Internet of Things Journal* 2 (6): 515–526.
19. Kodali, Ravi Kishore, Govinda Swamy, and Boppana Lakshmi. 2015. An implementation of IoT for healthcare. In *2015 IEEE recent advances in intelligent computational systems (RAICS)*, 411–416.
20. Farahani, Bahar, Farshad Firouzi, and Krishnendu Chakrabarty. 2020. Healthcare IoT. In *Intelligent internet of things: From device to fog and cloud*, 515–545.
21. Elhoseny, Mohamed, et al. 2018. Secure medical data transmission model for IoT-based healthcare systems. *IEEE Access* 6: 20596–20608.
22. Arijit, Ukil et al. 2016. IoT healthcare analytics: The importance of anomaly detection. In *2016 IEEE 30th international conference on advanced information networking and applications (AINA)*, 994–997. IEEE.
23. Zahid, Arnob et al. 2021. A systematic review of emerging information technologies for sustainable data-centric health-care. *International Journal of Medical Informatics* 149: 104420.
24. Khang, Alex. 2024. *AI and IoT technology and applications for smart healthcare systems*. CRC Press.
25. Tyagi, Amit Kumar and Richa. 2023. Digital Twin Technology: Opportunities and challenges for smart Era's applications. In *Proceedings of the 2023 fifteenth international conference on contemporary computing*, 328–336.
26. Khan, Osama et al. 2023. The future of pharmacy: How AI is revolutionizing the industry. *Intelligent Pharmacy* 1 (1): 32–40.
27. Haleem, Abid, et al. 2023. Exploring the revolution in healthcare systems through the applications of digital twin technology. *Biomedical Technology* 4: 28–38.
28. Katsoulakis, Evangelia et al. 2024. Digital twins for health: A scoping review. *NPJ Digital Medicine* 7 (1): 77.
29. Sun, Tianze, Xiwang He, and Zhonghai Li. 2023. Digital twin in healthcare: Recent updates and challenges. *Digital Health* 9: 20552076221149651.
30. Elkefi, Safa, and Onur Asan. 2022. Digital twins for managing health care systems: Rapid literature review. *Journal of Medical Internet Research* 24 (8): e37641.
31. Erol, Tolga, Arif Furkan Mendi, and Dilara DoÄan. 2020. The digital twin revolution in healthcare. In *2020 4th international symposium on multidisciplinary studies and innovative technologies (ISMSIT)*, vol. 2020, 1–7. IEEE.

32. Chen, Jiayuan et al. 2023. Networking architecture and key supporting technologies for human digital twin in personalized healthcare: A comprehensive survey. *IEEE Communications Surveys & Tutorials* 26 (1): 706–746.
33. Bjelland, Øystein., et al. 2022. Toward a digital twin for arthroscopic knee surgery: A systematic review. *IEEE Access* 10: 45029–45052.
34. Narigina, Marta, Andrejs Romanovs, and Rasa Bruzgiene. 2024. Digital twin technology in healthcare: A literature review. In *2024. IEEE 11th workshop on advances in information, electronic and electrical engineering (AIEEE)*, vol. 2024, 1–8. IEEE.
35. Kabir, Md Rafiul et al. 2025. Digital twins in healthcare IoT: A systematic review. *High-Confidence Computing*, 100340.
36. Xames, Md Doulotuzzaman, and Taylan G. Topcu. 2024. A systematic literature review of digital twin research for healthcare systems: Research trends, gaps, and realization challenges. *IEEE Access* 12: 4099–4126.
37. Mourtzis, Dimitris, et al. 2021. A smart IoT platform for oncology patient diagnosis based on AI: Towards the human digital twin. *Procedia CIRP* 104: 1686–1691.
38. Weerarathna, Induni Nayodhara et al. 2024. Leveraging digital twin technology to combat cardiovascular disease: A comprehensive review. In *2024 2nd DMIHER international conference on artificial intelligence in healthcare, education and industry (IDICAIEI)*, 1–6. IEEE.
39. Martinez-Velazquez, Roberto, Rogelio Gamez, and Abdulmotaleb El Saddik. 2019. Cardio twin: A digital twin of the human heart running on the edge. In *2019 IEEE international symposium on medical measurements and applications (MeMeA)*, vol. 2019, 1–6. IEEE.
40. Susilo, Monica E., et al. 2023. Systems-based digital twins to help characterize clinical dose–response and propose predictive biomarkers in a Phase I study of bispecific antibody, mosunetuzumab, in NHL. *Clinical and Translational Science* 16 (7): 1134–1148.
41. Li, Xinxiu et al. 2022. A dynamic single cell-based framework for digital twins to prioritize disease genes and drug targets. *Genome Medicine* 14 (1): 48.
42. Lu, Qiuchen et al. 2020. Digital twin-enabled anomaly detection for built asset monitoring in operation and maintenance. *Automation in Construction* 118: 103277.
43. Wong, Johnny Kwok Wai, Janet Ge, and Sean Xiangjian He. 2018. Digitisation in facilities management: A literature review and future research directions. *Automation in Construction* 92: 312–326.
44. Karakra, Abdallah et al. 2019. HospiT'Win: A predictive simulation-based digital twin for patients pathways in hospital. In *2019 IEEE EMBS international conference on biomedical & health informatics (BHI)*, 1–4. IEEE.
45. Karakra, Abdallah. 2018. Pervasive computing integrated discrete event simulation for a hospital digital twin. In *2018 IEEE/ACS 15th international conference on computer systems and applications (AICCSA)*, 1–6. IEEE.
46. Han, Yilong et al. 2023. Digital twinning for smart hospital operations: Framework and proof of concept. *Technology in Society* 74: 102317.
47. Aluvalu, Rajanikanth et al. 2023. The novel emergency hospital services for patients using digital twins. *Microprocessors and Microsystems* 98: 104794.
48. Gorelova, Anastasiia, Santiago Meliá, and Diana Gadzhimusieva. 2024. A discrete event simulation of patient flow in an assisted reproduction clinic with the integration of a smart health monitoring system. *IEEE Access*.
49. Jameil, Ahmed K., and Hamed Al-Raweshidy. 2024. Ai-enabled healthcare and enhanced computational resource management with digital twins into task offloading strategies. *IEEE Access*.
50. Kleinbeck, Constantin et al. 2024. Neural digital twins: Reconstructing complex medical environments for spatial planning in virtual reality. *International Journal of Computer Assisted Radiology and Surgery* 1–12.

51. Tai, Yonghang et al. 2022. Digital-Twin-Enabled IoMT system for surgical simulation using rAC-GAN. *IEEE Internet of Things Journal* 9 (21): 20918–20931.
52. Laaki, Heikki, Yoan Miche, and Kari Tammi. 2019. Prototyping a digital twin for real time remote control over mobile networks: Application of remote surgery. *IEEE Access* 7: 20325–20336.
53. Obaid, Daniel R., et al. 2019. Computer simulated "Virtual TAVR" to guide TAVR in the presence of a previous Starr-Edwards mitral prosthesis. *Journal of Cardiovascular Computed Tomography* 13 (1): 38–40.
54. Shu, Hongchao et al. 2023. Twin-S: A digital twin for skull base surgery. *International Journal of Computer Assisted Radiology and Surgery* 18 (6): 1077–1084.
55. Sartaj, Hassan, Shaukat Ali, and Julie Marie Gjøby. 2024. Uncertainty-aware environment simulation of medical devices digital twins. *Software and Systems Modeling* 1–27.
56. Ahmed, Imran, Misbah Ahmad, and Gwanggil Jeon. 2022. Integrating digital twins and deep learning for medical image analysis in the era of COVID-19. *Virtual Reality & Intelligent Hardware* 4 (4): 292–305.
57. Bersani, Marcello M., et al. 2022. Engineering of trust analysis-driven digital twins for a medical device. In *European conference on software architecture*, 467–482. Springer.
58. Kalozoumis, Panagiotis G., et al. 2022. Towards the development of a digital twin for endoscopic medical device testing. In *Digital twins for digital transformation: Innovation in industry*, 113–145. Springer.
59. Bethencourt, Loic, et al. 2021. Guiding measurement protocols of connected medical devices using digital twins: A statistical methodology applied to detecting and monitoring lymphedema. *IEEE Access* 9: 39444–39465.
60. Iqbal, Sheikh MA, et al. 2021. Advances in healthcare wearable devices. *NPJ Flexible Electronics* 5 (1): 9.
61. Khan, Wahid et al. 2014. Implantable medical devices. *Focal Controlled Drug Delivery* 33–59.
62. Chen, Junxin et al. 2023. Digital twin empowered wireless healthcare monitoring for smart home. *IEEE Journal on Selected Areas in Communications.*
63. Zhu, Zhengxu, and Ray Y. Zhong. 2025. A digital twin enabled wearable device for customized healthcare. *Digital Twin* 2: 17.
64. Yu, Feng et al. 2024. Intelligent wearable system with motion and emotion recognition based on digital twin technology. *IEEE Internet of Things Journal.*
65. Yang, Haochen, and Zhihao Jiang. 2024. Decision support for personalized therapy in implantable medical devices: A digital twin approach. *Expert Systems with Applications* 243: 122883.
66. Ghaempanah, Faezeh et al. 2024. Metaverse and its impact on medical education and health care system: A narrative review. *Health Science Reports* 7 (9): e70100.
67. Sartaj, Hassan, Shaukat Ali, and Julie Marie Gjøby. 2024. MeDeT: Medical device digital twins creation with few-shot meta-learning. *ACM Transactions on Software Engineering and Methodology.*
68. Vallée, Alexandre. 2024. Envisioning the future of personalized medicine: Role and realities of digital twins. *Journal of Medical Internet Research* 26: e50204.
69. Zhang, Qinran et al. 2022. A novel computational framework for integrating multidimensional data to enhance accuracy in predicting the prognosis of colorectal cancer. *MedComm–Future Medicine* 1 (2): e27.
70. Wu, Chengyue et al. 2022. MRI-based digital models forecast patient-specific treatment responses to neoadjuvant chemotherapy in triple-negative breast cancer. *Cancer Research* 82 (18): 3394–3404.

71. Silva, Tiago Castanheira, Michel Eppink, and Marcel Ottens. 2024. Digital twin in high throughput chromatographic process development for monoclonal antibodies. *Journal of Chromatography A* 1717: 464672.
72. Namasudra, Suyel. 2024. *IoT and ML for information management: A smart healthcare perspective*. Springer.
73. Creemers, Jeroen HA, and Johannes Textor. 2023. Leveraging mathematical models to improve the statistical robustness of cancer immunotherapy trials. *Current Opinion in Systems Biology* 40: 100540.
74. Sel, Kaan et al. 2025. Survey and perspective on verification, validation, and uncertainty quantification of digital twins for precision medicine. *NPJ Digital Medicine* 8 (1): 40.
75. Wang, Yue et al. 2024. TWIN-GPT: Digital twins for clinical trials via large language model. *ACM Transactions on Multimedia Computing, Communications and Applications*.
76. Herwig, Christoph, Ralf Pörtner, and Johannes Möller. 2021. *Digital twins: Tools and concepts for smart biomanufacturing*. Springer.
77. Goel, Neha, and Ravindra Kumar Yadav. 2024. *Internet of things enabled machine learning for biomedical applications*. CRC Press.
78. Maradapu Vera Venkata Sai, Akshita et al. 2024. Navigating the digital twin network landscape: A survey on architecture, applications, privacy and security. *High-Confidence Computing* 4 (4): 100269.
79. Ghubaish, Ali et al. 2020. Recent advances in the internet-of-medical-things (IoMT) systems security. *IEEE Internet of Things Journal* 8 (11): 8707–8718.
80. Zachos, Georgios, 2023. An IoT, IoMT security testbed for anomaly-based intrusion detection systems. In *2023 IFIP networking conference (IFIP networking)*, 1–6. IEEE.
81. Kabir, Md Rafiul, and Sandip Ray. 2024. DT-IoMT: A digital twin reference model for secure internet of medical things. In *2024 IEEE computer society annual symposium on VLSI (ISVLSI)*, 433–438. IEEE.
82. Zhang, Jun, et al. 2020. Cyber resilience in healthcare digital twin on lung cancer. *IEEE Access* 8: 201900–201913.
83. Jimenez, Jaime Ibarra, Hamid Jahankhani, and Stefan Kendzierskyj. 2020. Health care in the cyberspace: Medical cyber-physical system and digital twin challenges. In *Digital twin technologies and smart cities*, 79–92.
84. Sharma, Vikas, Akshi Kumar, and Kapil Sharma. 2024. Digital twin: Securing IoT networks using integrated ECC with blockchain for healthcare ecosystem. *Knowledge and Information Systems* 1–32.
85. Lutze, Rainer. 2019. Digital twins in ehealth–: Prospects and challenges focussing on information management. In *2019 IEEE international conference on engineering, technology and innovation (ICE/ITMC)*, 1–9. IEEE.
86. Sadman Sakib Akash and Md Sadek Ferdous. 2022. A blockchain based system for healthcare digital twin. *IEEE Access* 10: 50523–50547.
87. Zheng, Yandong, 2021. Towards private similarity query based healthcare monitoring over digital twin cloud platform. In *2021 IEEE/ACM 29th international symposium on quality of service (IWQOS)*, 1–10. IEEE.
88. Stephanie, Veronika, Ibrahim Khalil, and Mohammed Atiquzzaman. 2023. Digital twin enabled asynchronous SplitFed learning in E-healthcare systems. *IEEE Journal on Selected Areas in Communications* 41 (11): 3650–3661.
89. Tang, Yongyi et al. 2023. Secure data sharing and prediction with digital twin and blockchain in healthcare. *IEEE Communications Magazine*.
90. De Benedictis, Alessandra et al. 2022. Digital twins in healthcare: an architectural proposal and its application in a social distancing case study. *IEEE Journal of Biomedical and Health Informatics* 27 (10): 5143–5154.

91. Iqbal, Mubashar et al. 2023. Towards healthcare digital twin architecture. In *International conference on business informatics research*, 45–60. Springer.
92. Pellegrino, G., M. Gervasi, and M. Angelelli. 2024. A conceptual framework for digital twin in healthcare: Evidence from a systematic meta-review. *Information Systems*.
93. Alqahtani, Abdullah, Shtwai Alsubai, and Munish Bhatia. 2023. Digital-twin-assisted healthcare framework for adult. *IEEE Internet of Things Journal* 11 (8): 14963–14970.
94. Laamarti, Fedwa, et al. 2020. An ISO/IEEE 11073 standardized digital twin framework for health and well-being in smart cities. *IEEE Access* 8: 105950–105961.
95. Suraci, Chiara, et al. 2022. The next generation of eHealth: A multidisciplinary survey. *IEEE Access* 10: 134623–134646.
96. Liu, Ying, et al. 2019. A novel cloud-based framework for the elderly healthcare services using digital twin. *IEEE Access* 7: 49088–49101.
97. Garg, Hitendra et al. 2022. Spoofing detection system for e-health digital twin using EfficientNet convolution neural network. *Multimedia Tools and Applications* 81 (19): 26873–26888.
98. Lutze, Rainer. 2020. Digital twin based software design in eHealth-a new development approach for health/medical software products. In *2020 IEEE international conference on engineering, technology and innovation (ICE/ITMC)*, 1–9. IEEE.
99. Mila, Sumaiya Afroz et al. 2025. MASC: Wearable design for infectious disease detection through machine learning. *IEEE Access*.
100. Akram, Junaid et al. 2024. Ai-generated content-as-a-service in IOMT-based smart homes: Personalizing patient care with human digital twins. *IEEE Transactions on Consumer Electronics*.
101. Shoukat, Muhammad Usman et al. 2024. Smart home for enhanced healthcare: Exploring human machine interface oriented digital twin model. *Multimedia Tools and Applications* 83 (11): 31297–31315.
102. Rivera, Luis F. et al. 2019. Towards continuous monitoring in personalized healthcare through digital twins. In *Proceedings of the 29th annual international conference on computer science and software engineering*, 329–335.
103. Brahmi, Rafika, Noureddine Boujnah, and Ridha Ejbali. 2024. Elaboration of innovative digital twin models for healthcare monitoring with 6g functionalities. *IEEE Access* 12.
104. Tao, Fei et al. 2019. Digital twin-driven product design framework. *International Journal of Production Research* 57 (12): 3935–3953.
105. Ahmad, Muhammad Aurangzeb et al. 2023. Validation of a hospital digital twin with machine learning. In *2023 IEEE 11th international conference on healthcare informatics (ICHI)*, 465–469. IEEE.
106. Alsubai, Shtwai et al. 2023. Hybrid IoT-edge-cloud computing-based athlete healthcare framework: Digital twin initiative. *Mobile Networks and Applications* 1–20.
107. Lv, Zhihan, Jinkang Guo, and Haibin Lv. 2023. Deep learning-empowered clinical big data analytics in healthcare digital twins. *IEEE/ACM Transactions on Computational Biology and Bioinformatics*.
108. Iqram Hussain, Md., Azam Hossain, and Se-Jin. Park, 2021. A healthcare digital twin for diagnosis of stroke. In *2021, IEEE international conference on biomedical engineering, computer and information technology for health (BECITHCON)*, 18–21. IEEE.
109. Chen, Jiayuan et al. 2024. A revolution of personalized healthcare: Enabling human digital twin with mobile AIGC. *IEEE Network*.
110. Gebreab, Senay et al. 2024. Accelerating digital twin development with generative AI: A framework for 3D modeling and data integration. *IEEE Access*.
111. Khan, Sagheer, et al. 2023. A Novel Digital Twin (DT) model based on WiFi CSI, signal processing and machine learning for patient respiration monitoring and decision-support. *IEEE Access* 11: 103554–103568.

112. Sartaj, Hassan et al. 2024. Model-based digital twins of medicine dispensers for healthcare IoT applications. *Software: Practice and Experience* 54 (6): 1172–1192.
113. Amudhavalli, P., et al. 2025. Investigating the revolution of healthcare application with intense comparisons and case study. In *The impact of algorithmic technologies on healthcare*, 421–445.
114. Wang, Chenyu, Zhipeng Cai, and Yingshu Li. 2022. Sustainable blockchain-based digital twin management architecture for IoT devices. *IEEE Internet of Things Journal* 10 (8): 6535–6548.
115. Suhail, Sabah et al. 2021. Trustworthy digital twins in the industrial internet of things with blockchain. *IEEE Internet Computing* 26 (3): 58–67.
116. Zhu, Saide et al. 2018. Coin hopping attack in blockchain-based IoT. *IEEE Internet of Things Journal* 6 (3): 4614–4626.
117. Boopathi Raja, G., et al. 2025. Collaboration of blockchain-based digital twins in the healthcare industry: From idea to implementation. In *Blockchain-based digital twins: Research trends and challenges*, 223.
118. Manocha, A., Y. Afaq, and M. Bhatia. 2023. Digital Twin-assisted Blockchain-inspired irregular event analysis for eldercare. *Knowledge-Based Systems*.
119. Kumari, D., P. Kumar, and S. Prajapat. 2024. A blockchain assisted public auditing scheme for cloud-based digital twin healthcare services. *Cluster Computing*.
120. Sheng, Bo et al. 2023. Detecting latent topics and trends of digital twins in healthcare: A structural topic model-based systematic review. *Digital Health* 9: 20552076231203672.
121. Chen, Jiayuan et al. 2024. Generative AI-driven human digital twin in IoT-healthcare: A comprehensive survey. *IEEE Internet of Things Journal*.
122. Venkatesh, Kaushik P., Marium M. Raza, and Joseph C. Kvedar. 2022. Health digital twins as tools for precision medicine: Considerations for computation, implementation, and regulation. *NPJ Digital Medicine* 5 (1): 150.
123. Dolin, Robert H., et al. (2001). The HL7 clinical document architecture. *Journal of the American Medical Informatics Association* 8 (6): 552–569.
124. Makadia, Rupa, and Patrick B Ryan. 2014. Transforming the premier perspective hospital database into the observational medical outcomes partnership (omop) common data model. *Egems* 2 (1).
125. Drummond, David, and Apolline Gonsard. 2024. Definitions and characteristics of patient digital twins being developed for clinical use: Scoping review. *Journal of Medical Internet Research* 26: e58504.
126. De Domenico, Manlio et al. 2025. Challenges and opportunities for digital twins in precision medicine from a complex systems perspective. *NPJ Digital Medicine* 8 (1): 37.

Digital Twin Tools for Smart Manufacturing 5

Digital twin tools shaping the evolution of intelligent manufacturing in Industry 4.0

5.1 Introduction

Industry 4.0 represents a transformative era characterized by the convergence of cyber-physical systems, pervasive connectivity through the Industrial Internet of Things (IIoT), advanced analytics, and increased automation in manufacturing environments [1, 2]. At the core of this revolution lies the digital twin technology, that creates high-fidelity digital replicas of physical assets, processes, and systems [3]. Unlike conventional static simulation models, digital twins maintain dynamic, real-time synchronization with their physical counterparts, enabling seamless cyber-physical integration [4]. This capability supports manufacturers in achieving real-time monitoring, predictive insights, and proactive optimization, fundamentally reshaping traditional manufacturing paradigms [5].

Digital twins represent a significant paradigm shift, transitioning manufacturing from a reactive, historically data-driven approach toward predictive, autonomous, and resilient operational strategies. This shift is crucial as manufacturers increasingly grapple with complexity in global supply chains, sustainability requirements, and demand for rapid product innovation [6]. Modern digital twin tools and platforms leverage extensive IIoT integration, cloud computing, and advanced visualization technologies, facilitating comprehensive lifecycle management and optimized resource allocation across complex manufacturing ecosystems [7]. A pivotal component accelerating the effectiveness and adoption of digital twins is their integration with artificial intelligence (AI) and machine learning (ML) technologies [8]. AI and ML significantly enhance DT capabilities, providing powerful tools for anomaly detection, predictive maintenance, and real-time decision-making processes [9].

M. R. Kabir and S. Ray, *Digital Twins for Distributed IoT Applications*, Synthesis Lectures on Engineering, Science, and Technology, https://doi.org/10.1007/978-3-032-18938-7_5

More recently, large language models (LLMs) have emerged as a novel means of interacting with digital twin environments. LLMs facilitate natural language-based querying, reporting, and scenario analysis, greatly enhancing accessibility and interpretability of complex DT data for decision-makers [10]. By embedding these intelligent methodologies, digital twins transcend basic monitoring functionalities and evolve into fully interactive and autonomous systems capable of dynamic adaptation and learning.

Given the rapid evolution and growing complexity of DT technologies, there is a need for practical guidance on assembling production-grade toolchains and aligning them with emerging standards [11]. This chapter critically reviews state-of-the-art DT tools for manufacturing-mapping strengths, limitations, and adoption patterns. We examine how the ISO 23247 framework supports interoperable, scalable twins while identifying practical challenges and near-term opportunities. We illustrate these points with a semiconductor manufacturing case study, which illustrates how multi-layer twins integrate with AI/ML to deliver early yield prediction and closed-loop recipe optimization. The insights provided through this perspective are valuable for researchers, practitioners, and industry stakeholders seeking to effectively use DT technologies within modern manufacturing settings. Highlighting both current capabilities and future opportunities, this chapter underscores the strategic importance of digital twins as critical enablers of Industry 4.0 and beyond.

5.2 Digital Twin in Manufacturing: Concepts and Role

5.2.1 Industry 4.0, IIoT and Beyond

Industry 4.0 represents the broader transformation of manufacturing through advanced digital technologies, often referred to as the "Fourth Industrial Revolution" [12]. It follows previous waves of industrial change-mechanization, mass production, and automation-and aims to create smart, autonomous, and highly efficient production systems. At the heart of Industry 4.0 are technologies like IIoT, cyber-physical systems, digital twins, artificial intelligence, cloud computing, 3D printing, and big data analytics [13].

Among these, IIoT plays a foundational role by connecting industrial devices such as machines, sensors, and controllers, enabling them to communicate and share data in real time. Unlike traditional industrial automation, IIoT brings a new level of connectivity and intelligence by collecting vast amounts of real-time data from physical assets and making it accessible for analysis and decision-making. When integrated with DTs, IIoT allows for continuous monitoring and deeper insights into system behavior, enabling predictive maintenance, process optimization, and improved product quality. For manufacturers, this means fewer unexpected breakdowns, more efficient use of resources, and faster response to changing conditions. Ultimately, IIoT serves as the digital backbone of smart manufacturing, helping industries move toward more agile, responsive, and data-driven operations.

5.2.2 Digital Twin Manufacturing Market

To better understand the growing relevance and impact of digital twin technology in smart manufacturing, it is useful to examine key market trends, investment dynamics, and technological adoption patterns. Figure 5.1 illustrates the explosive growth of the global digital twins market within the manufacturing sector. Valued at USD 10.27 billion in 2023, the market is projected to reach approximately USD 714.01 billion by 2032, driven by a staggering compound annual growth rate (CAGR) of 60.20% [14]. The exponential rise reflects increasing adoption across industries to enhance operational efficiency, product quality, and predictive capabilities. The chart clearly demonstrates a shift from early adoption to widespread deployment as DTs become foundational to smart manufacturing strategies.

Figure 5.2 compares the estimated implementation costs with the projected benefits, using market size as a proxy, over a 10-year span. While implementation costs increase steadily (e.g., from USD 8 billion in 2023 to USD 30 billion in 2032), the associated benefits grow exponentially, mirroring the market value trend. This highlights a strong return on investment (ROI) for early adopters, especially as the technology matures and economies of scale reduce deployment costs.

The trend line in Fig. 5.3 shows the percentage of manufacturers integrating IIoT with DTs over time. Starting at just 10% in 2023, the adoption is expected to surpass 90% by 2032, reflecting a rapidly growing reliance on real-time data connectivity and sensor-driven analytics.[1] This integration is crucial for enabling continuous monitoring, predictive maintenance, and system optimization-core pillars of Industry 4.0.

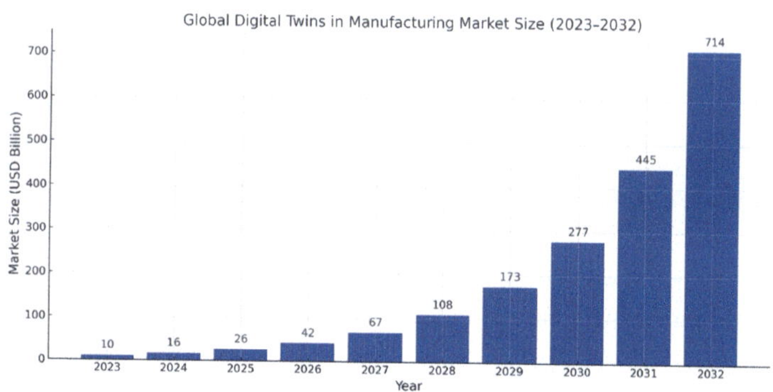

Fig. 5.1 Global Digital Twins in Manufacturing Market Size (2023–2032). The market is projected to grow from USD 10.27 billion in 2023 to USD 714.01 billion in 2032, reflecting a compound annual growth rate (CAGR) of 60.20%

[1] Figures 5.2 and 5.3 are produced from the authors' own analysis based on aggregated market data found from source [14].

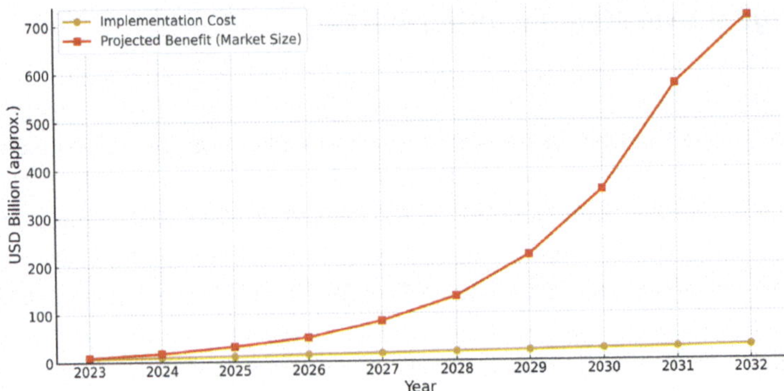

Fig. 5.2 Cost versus Benefit of Digital Twin in Manufacturing. While implementation costs increase steadily, the benefits, represented by market value, grow exponentially, indicating a strong return on investment

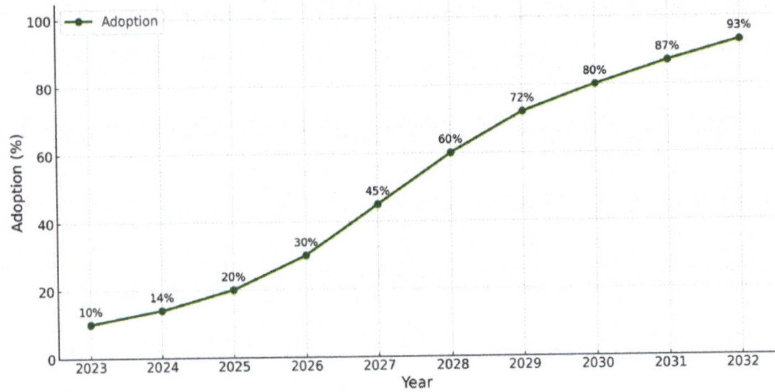

Fig. 5.3 IIoT–Digital twin integration adoption (2023–2032)

5.3 Commercial Digital Twin Tools and Software Platforms

To bring order to the crowded digital-twin landscape, we group the tools most commonly encountered in smart-manufacturing projects into three functional categories (i.e., C1, C2, and C3) that mirror the typical life cycle of a factory asset. The taxonomy is shown in Fig. 5.4.

To ensure transparency in the review scope, the selection of commercial DT tools followed a structured set of criteria. Tools were included based on (i) publicly documented industrial adoption of at least ten deployments or equivalent case studies, (ii) availability of developer APIs or open integration interfaces supporting IIoT and simulation workflows, and (iii) relevance to recent manufacturing applications reported. Bibliographic sources were gathered from IEEE Xplore and ScienceDirect using search terms such as "digital twin manufacturing," and "digital twin commercial tools" along with the tool names.

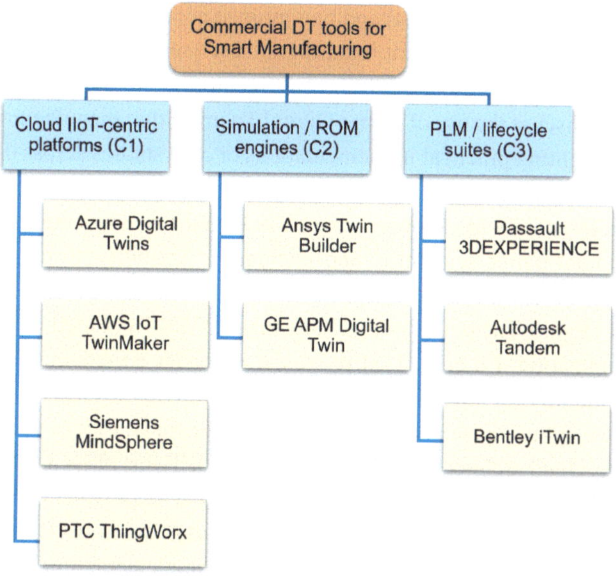

Fig. 5.4 Taxonomy of commercial DT tools for smart-manufacturing in functional categories

5.3.1 Cloud IIoT-Centric Platforms

Cloud IIoT-centric platforms serve as the backbone of scalable DT ecosystems by integrating IIoT data with cloud computing capabilities. These platforms enable real-time data acquisition, edge-to-cloud processing, and remote monitoring of physical assets, allowing manufacturers to visualize, analyze, and optimize operations from anywhere. By providing built-in support for device connectivity, time-series databases, and interactive dashboards, platforms in this category (e.g., AWS IoT TwinMaker and Azure Digital Twins) empower smart manufacturing with agility and responsiveness.

5.3.1.1 Microsoft Azure Digital Twins

Microsoft Azure Digital Twins [15] is a fully managed platform-as-a-service that enables the construction of rich, live digital replicas of physical environments-ranging from individual machines to entire smart cities. Using the open Digital Twins Definition Language (DTDL), users define entities and their relationships, which are then stored as a graph that can be queried in real time. The platform integrates seamlessly with Azure IoT Hub and other data sources, allowing telemetry to stream directly into the digital twin graph, where event routing, Time Series Insights, Functions, and AI tools can be employed for analysis or automated decision-making. With support for multiple development environments (i.e., C#, Python, Java, and JavaScript) and fine-grained access control, Azure Digital Twins facilitates the development of secure, data-driven DT solutions across domains such as manufacturing, smart buildings, and urban infrastructure.

Several studies discuss the use of Azure Digital Twins within manufacturing-related scenarios, often highlighting practical implementations or case studies across different sectors. For example, Parle et al. [16] performed a comparative analysis of several DT platforms, including Azure Digital Twins, emphasizing its potential for achieving net-zero manufacturing through cloud-native scalability and integration flexibility. Similarly, Fahim et al. [17] presented a practical demonstration of a wind turbine DT model utilizing Azure's IoT capabilities and Digital Twins Definition Language (DTDL) for predictive modeling and operational optimization. Lu et al. [18] provided a broader review of DT applications in smart factories, identifying cloud services such as Azure Digital Twins as key enablers for scalable, real-time cyber-physical integration. In another study, Picone et al. [19] proposed a flexible edge-to-cloud architecture where Azure Digital Twins plays a central role in managing distributed twin instances and facilitating modular deployment strategies. Finally, Lehner et al. [20] explored model-driven engineering approaches for DTs, explicitly mentioning Azure Digital Twins Explorer as a reference implementation for managing data models generated from system transformations.

5.3.1.2 Siemens MindSphere (Insights Hub)

As part of Siemens' Xcelerator portfolio, NX and MindSphere (now known as Insights Hub) form a powerful ecosystem for implementing closed-loop DT solutions. Siemens NX is a comprehensive, cloud-enabled CAD/CAM/CAE platform that supports generative design, simulation, and manufacturing planning-all within a single, model-driven environment. MindSphere extends this digital workflow into the physical world by connecting industrial assets via IIoT. It collects real-time data from sensors and machines, either at the edge or in the cloud, and applies analytics to monitor performance, detect anomalies, and drive improvements. Together, NX creates the virtual model, while MindSphere brings in real-world data, enabling a continuous feedback loop. This integration helps manufacturers accelerate design iterations, improve quality, and develop new service-oriented business models.

A case study [21] of Siemens MindSphere explored how boundary resources (such as APIs and SDKs) facilitate the development of IIoT ecosystems by enabling third-party integration and innovation. The study highlights that MindSphere's openness and provision of technical resources support a growing network of partners, driving platform-based end-to-end industrial solutions. This work extends the theoretical understanding of boundary resources in the B2B domain, emphasizing their critical role in shaping ecosystem emergence and value creation within IIoT platforms.

5.3.1.3 AWS IoT TwinMaker

AWS IoT TwinMaker [22] provides a distinctive approach to DT development by seamlessly integrating disparate data sources-including IoT sensor streams, video feeds, and enterprise databases-into cohesive, real-time 3D visualizations. Its flexible architecture enables developers and engineers to construct detailed virtual representations without relocating underlying data, thereby enhancing scalability and responsiveness in industrial monitoring and control tasks. By facilitating real-time insight and predictive analytics, AWS IoT TwinMaker plays a transformative role in sectors ranging from manufacturing facilities to intelligent infrastructure management. Figure 5.5 shows a generic architecture for the AWS IoT TwinMaker approach.

5.3.1.4 PTC ThingWorx

PTC ThingWorx is more than just another IoT platform-it is a powerful engine for bringing DTs to life. Its robust integration capabilities with sensors, control systems, and analytical tools allow precise modeling of complex industrial processes and products. ThingWorx's ability to deliver real-time predictive insights significantly enhances operational decision-making, promoting efficiency and proactive maintenance strategies across industries such as automotive, aerospace, and industrial manufacturing [23]. The majority of its applications have been in IIoT factory optimization and platform development [24, 25], but its use in creating DTs is still in its infancy. For instance, Bellalouna et al. [26] relied on estimated loading data and high safety factors during product design, which often leads to over-dimensioning and inefficient material use. They aimed to apply the DT approach to accurately capture real operational loads, enabling more precise and sustainable load-oriented product design.

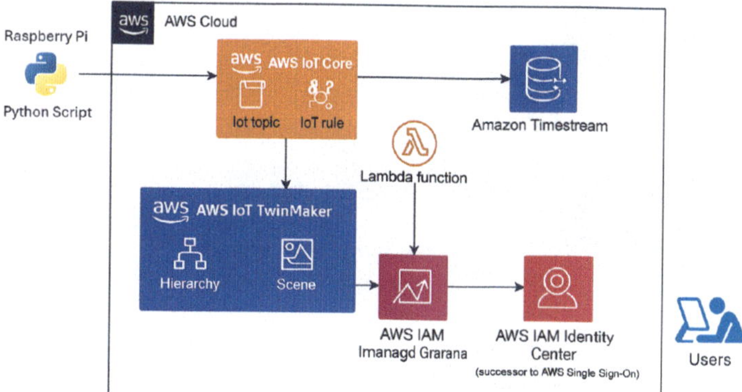

Fig. 5.5 High-level architecture of AWS IoT TwinMaker solution

5.3.2 High-Fidelity Simulation Engines

High-fidelity simulation engines specialize in creating accurate virtual replicas of physical systems through advanced physics-based modeling and numerical methods. These tools support multi-domain simulations to replicate mechanical, electrical, thermal, and control dynamics with high precision. By enabling scenario testing, predictive maintenance, and system optimization, they play a critical role in the design and validation of complex manufacturing systems before deployment.

5.3.2.1 Ansys Twin Builder

Ansys Twin Builder [27] is a powerful engineering simulation platform designed to create and deploy high-fidelity DTs of physical assets. By integrating real-world data with physics-based modeling and simulation, Twin Builder enables engineers to design, validate, and optimize systems throughout their operational lifecycle. It supports seamless co-simulation with various tools and allows for real-time monitoring, predictive maintenance, and performance optimization. With its strong foundation in multiphysics modeling and system-level simulation, Ansys Twin Builder accelerates the development of robust DTs for industries such as aerospace, automotive, energy, and manufacturing. Figure 5.6 depicts a workflow for hybrid DT deployment by Ansys Twin Builder.

A couple of studies [28, 29] highlighted the use of ANSYS Twin Builder for induction motor control. One developed a vector Field Oriented Control (FOC) system using reduced-order modeling and coupled simulations with ANSYS Maxwell. The other focused on stabilizing a synchronous generator's rotor speed via scalar control of an induction motor using a PID-regulated inverter. Both studies showcased Twin Builder's ability to simulate

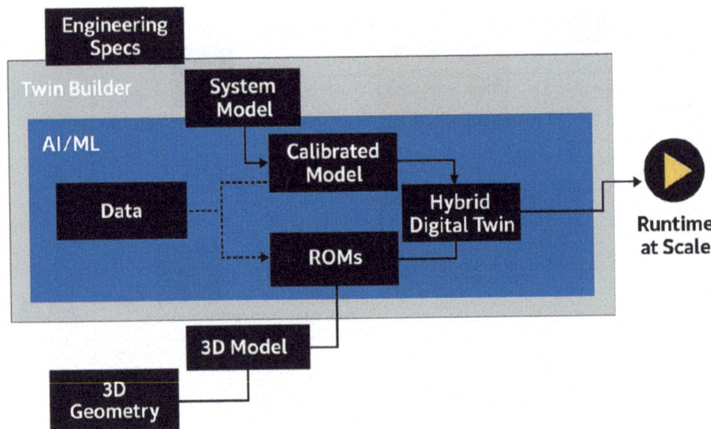

Fig. 5.6 Ansys Twin Builder workflow for hybrid digital twin deployment

high-fidelity motor control systems using built-in libraries and imported field-based models. Another work [30] focused on a synchronous generator with permanent magnets in a wind power plant and explored two methods for stabilizing its rotational speed: one using a PID-controlled electromagnetic clutch and the other using a squirrel-cage induction motor controlled by scalar law via an inverter. The complete energy conversion and transmission process is simulated using ANSYS Electromagnetics, with the generator modeled in ANSYS Maxwell and integrated into Twin Builder for co-simulation.

5.3.2.2 GE Vernova Predix

GE Vernova's Predix [31] stands apart as a DT platform tailored specifically for heavy industries aiming for sustainability and efficiency. By embedding advanced analytics and machine learning into its core, Predix provides organizations with proactive visibility into their assets—spotting operational risks and inefficiencies before they occur. From reducing energy consumption in industrial plants to ensuring reliability in renewable energy operations, Predix is the digital backbone for transformative, predictive asset management.

5.3.3 PLM and Lifecycle-Driven Virtual Twins

Product Lifecycle Management (PLM) and lifecycle-driven platforms focus on maintaining digital continuity across a product's lifecycle-from conceptual design and engineering to manufacturing and maintenance. Tools like Dassault 3DEXPERIENCE, Autodesk Tandem, and Bentley iTwin provide robust data management, collaboration features, and traceability, ensuring that the DT evolves in sync with its physical counterpart.

5.3.3.1 Dassault 3DEXPERIENCE

Dassault Systèmes' 3DEXPERIENCE platform uniquely integrates sophisticated DT modeling with collaborative innovation tools, effectively bridging virtual simulations with physical operations. The platform's extensive simulation and visualization capabilities enable engineers and stakeholders to collaboratively design, validate, and optimize complex products and systems in a virtual environment. Widely adopted in automotive, aerospace, and smart manufacturing domains, the 3DEXPERIENCE platform accelerates sustainable innovation by streamlining decision-making processes and enhancing resource efficiency [32].

5.3.3.2 Bentley iTwin Platform

Bentley System's iTwin Platform [33] provides a specialized cloud-based infrastructure DT solution, which converts static engineering data into dynamic, interactive DTs of physical assets. Its open architecture, complemented by comprehensive APIs, facilitates real-time collaboration, lifecycle asset management, and seamless integration of multidisciplinary data sources. The iTwin Platform significantly advances infrastructure project planning, construction, and operational management, offering transformative efficiency and enhanced transparency for urban development and infrastructure projects. Parle et al. [16] presented a comprehensive case study demonstrating the application of the Bentley iTwin platform in a manufacturing context. They developed a DT framework that integrates real-time sensor data with engineering models to monitor and optimize manufacturing processes. This approach facilitated enhanced decision-making, predictive maintenance, and operational efficiency within the manufacturing environment.

5.3.3.3 Autodesk Tandem

Autodesk Tandem [34] introduces a distinct BIM-centric approach to DT technology, enabling detailed virtual representations of built environments. By integrating static BIM data into dynamic operational models, Tandem empowers facility managers and engineers to monitor, manage, and optimize building performance effectively. This approach transforms facility management practices by providing actionable insights throughout the building lifecycle, thereby improving operational efficiency and sustainability outcomes in commercial and industrial properties. Autodesk Tandem has been used primarily in other domains (e.g., energy management, healthcare facility, smart city, etc.). In the case of manufacturing, Elias et al. [35] proposed using DTs to enhance facilities and asset management by transforming static BIM data into dynamic, actionable models. They demonstrated this approach through Autodesk Tandem, which is designed to support real-time collaboration and data integration in the AECO industry.

5.3.4 Comparative Analysis

To contextualize the diverse roles DT platforms play in smart manufacturing, we present a comparative analysis of their industrial adoption and typical applications. Table 5.1 categorizes leading tools based on their core functionality-ranging from cloud-based IIoT platforms (C1) and simulation-driven engines (C2) to product lifecycle management (PLM) suites (C3). For each platform, we report publicly disclosed customer counts or flagship deployments, outline representative use cases, and limitations that illustrate how these tools are deployed on factory floors. This overview highlights not only the maturity and scope of each solution but also the varying depths of integration with production systems, simulation tools, and enterprise architectures.

Table 5.1 Commercial Digital Twin tools in manufacturing: public adoption, typical use cases, and limitations

Category	Platform	Typical manufacturing use-case	Limitations
C1	Siemens MindSphere	Cloud + edge IIoT PaaS, asset monitoring, predictive maintenance, bridging to Siemens automation stack	Siemens-centric; interoperability with non-Siemens hardware may require extra integration
C1	AWS IoT TwinMaker	Cloud-native, low-code twins combining SiteWise data, 3D/BIM, and video for asset dashboards, connected-worker apps, and production monitoring [36]	Early maturity, dependent on AWS ecosystem, limited built-in simulation support
C1	PTC ThingWorx + Kepware Edge	Production performance roll-ups, KPI boards, and predictive maintenance initiatives-particularly in plants utilizing Kepware PLC connectors or Rockwell/Dell validated edge stacks.	Reliance on PTC/Kepware stacks for optimal integration, moderate learning curve for advanced setups
C1	Microsoft Azure Digital Twins	Hyperscale PaaS for twin graphs, event routing, AI at scale; flexible when paired with Ansys, PTC, Unity, etc.	Cloud-dependent, complex setup, limited built-in physics modeling
C2	Ansys Twin Builder	Physics-based hybrid twins, ROMs, HIL/SIL validation, heavy CAE co-simulation	Requires advanced simulation knowledge, limited real-time IoT analytics capabilities without external platforms
C2	GE Vernova Predix	Hybrid physics + ML twins for rotating equipment (turbines, compressors, generators), driving maintenance and optimization across multi-site fleets.	Narrow industry focus (heavy industry, energy), high complexity for setup and maintenance
C3	Dassault Systèmes 3DEXPERIENCE	End-to-end product lifecycle, immersive "virtual twin" factories, quality planning, MES/ERP integration	Heavyweight, high-complexity solution with expensive deployments, best suited for enterprises over SMEs
C3	Bentley iTwin	Federated "evergreen" twins of factories, plants, and logistics sites-tracking change across CAD/BIM, reality-capture scans, and IoT to coordinate design-build-operate workflows [37]	Focused heavily on infrastructure and civil engineering, limited IoT analytics capabilities
C3	Autodesk Tandem	Data-rich handover twins for new or retro-fitted industrial buildings-tagging equipment, linking O&M manuals, and visualizing space utilization for facilities teams.	Still early maturity in industrial contexts, limited built-in simulation and predictive analytics

5.4 Generalized Digital Twin Development Tools

5.4.1 Mathematical Modeling and Simulation

MATLAB/Simulink

MATLAB/Simulink is a popular software tool used by engineers to build, simulate, and analyze dynamic systems. Its easy-to-use graphical interface allows users to quickly create digital models that closely mimic physical systems, making it highly suitable for developing DTs. By connecting sensor data, implementing control algorithms, and performing real-time predictions, MATLAB/Simulink effectively bridges the gap between virtual simulations and real-world applications. With specialized toolboxes for IoT integration, machine learning, and data analytics, it has become an essential platform for engineers developing DTs in fields like manufacturing, automotive, aerospace, and healthcare.

Kombaya et al. [38] presented a DT design and simulation model for reconfigurable manufacturing systems (RMSs) using MATLAB and Simulink. Their approach combines data-driven and physics-based modeling to enable prediction, anomaly detection, and hypothesis testing within a dynamic, multi-station manufacturing setup. By leveraging tools like SimEvents and Stateflow, they simulate both continuous and discrete behaviors, aiming to automate system reconfiguration and improve adaptability to changing production demands. Cimino et al. [39] explained how MATLAB/Simulink enables the development of high-fidelity DTs in manufacturing by combining physics-based modeling, real-time data integration, and predictive analytics. Simulink is used to replicate machine behavior, while tools like Stateflow and SimEvents simulate control logic and discrete workflows. Data-driven toolboxes assist in fault detection, model validation, and system optimization, making the platform well-suited for smart, reconfigurable manufacturing systems.

Python Scientific Libraries

Python scientific libraries (i.e., *NumPy, SciPy, and SimPy*) provide a versatile, open-source framework for developing DTs in manufacturing and industrial applications. *NumPy* enables efficient numerical computation and large-scale data handling essential for real-time data processing, while *SciPy* offers advanced mathematical algorithms for optimization, signal processing, and system modeling. *SimPy* complements these capabilities by facilitating discrete-event simulations, which are crucial for modeling manufacturing workflows, scheduling, and logistics. Together, these libraries create a flexible and accessible ecosystem, allowing developers and engineers to rapidly prototype and implement customized DT solutions tailored to specific industrial challenges. For example, Mattera et al. [40] proposed a reinforcement learning (RL) workflow for industrial manufacturing control, specifically applied to a wire arc additive manufacturing (WAAM) process, where a process-based reward function, Reduced Order Model, and Deep Deterministic Policy Gradient (DDPG) controller were developed. Using NumPy for numerical computation, they integrated a realistic WAAM simulator to enable sim-to-real transfer and automatic code generation for deployment on the motion platform controller.

5.4.2 AI, ML, and Natural Language Interaction

ML Frameworks

Industrial machine learning (I-ML) is increasingly central to Industry 4.0, enabling predictive maintenance, anomaly detection, and quality control while facing challenges of reproducibility, scalability, and reliability. Against this backdrop, the following frameworks illustrate how such capabilities are being realized within DTs [41]. Machine Learning (ML) frameworks such as TensorFlow [42], PyTorch [43], and scikit-learn [44] are increasingly critical in the development of intelligent DTs in smart manufacturing. TensorFlow and PyTorch enable deep learning-based predictive modeling, anomaly detection, and dynamic optimization through neural network architectures, while scikit-learn provides robust algorithms for classical machine learning tasks like clustering, regression, and classification [45, 46]. Perno et al. [47] presented a framework for developing machine learning-based DTs to predict critical process parameters in real time within the process manufacturing industry. The framework was validated through a case study at an international manufacturing company, demonstrating improved prediction accuracy and highlighting both the potential benefits and challenges of implementing DT–ML systems in industrial settings. Furthermore, multiple studies [48–51] extensively discussed the role of DT within ML, deep learning, and comprehensive frameworks.

Large Language Models

Large Language Models (LLMs) such as ChatGPT and GPT-4 present transformative potential for DT applications by enabling advanced natural language interaction, knowledge integration, and reasoning capabilities. These generative AI models facilitate intuitive human-machine collaboration, help interpret complex data insights, and automate report generation and decision recommendations. Recent studies [10, 52] highlighted their capacity to improve the usability of DTs, particularly in contexts where non-expert stakeholders must understand sophisticated manufacturing insights, perform guided troubleshooting, or conduct scenario analyses through conversational interfaces.

Twin Talk GPT, by EOT (Embassy of Things) [53], is a generative AI solution designed to enhance DTs at the industrial edge. It combines advanced AI models with EOT's edge controllers, enabling real-time simulation and analytics of operational data for predictive maintenance, anomaly detection, and process optimization. Integrated within EOT's Industrial Data Fabric, Twin Talk GPT supports secure and seamless data exchange between

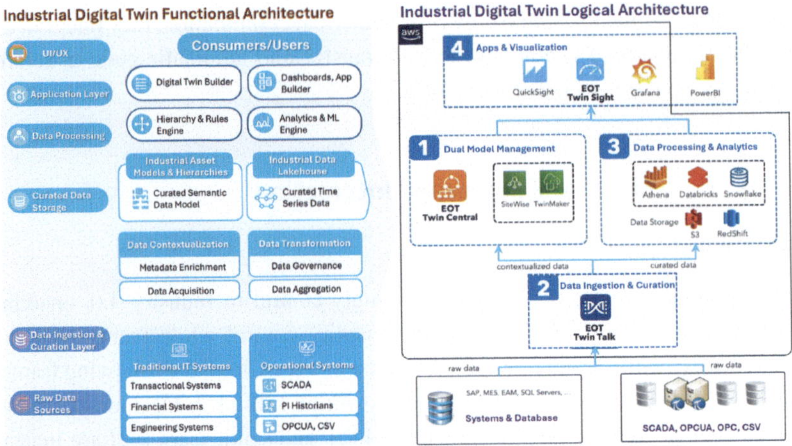

Fig. 5.7 Twin Talk GPT conceptual architecture for digital twin

industrial equipment, control systems, and analytics platforms, making industrial processes smarter, more responsive, and easier to manage. A conceptual architecture for an industrial DT by Twin Talk GPT is depicted in Fig. 5.7.

Cloud AI Platforms

Cloud-based AI platforms (e.g., Azure AI, Amazon SageMaker, Google AI, etc.) are reshaping how manufacturers build and manage DTs by offering powerful, flexible tools that go beyond traditional software capabilities. With built-in tools for training sophisticated machine learning models, deploying them quickly at scale, and running real-time analytics, these platforms streamline the integration of AI into everyday operations. This means manufacturers can swiftly turn massive amounts of data into actionable insights, monitor their operations globally in real time, and anticipate problems before they occur-making their processes smarter, more adaptive, and more competitive on a global stage [3, 18].

5.5 ISO 23247 Standard for Digital Twin in Manufacturing

5.5.1 Role and Significance in Current Practices

ISO 23247 is an emerging international standard that defines a DT framework for manufacturing, providing common terminology and a multi-layer reference architecture for implementing DTs on the factory floor. It partitions a DT system into distinct layers and entities, from observable manufacturing elements (e.g., personnel, equipment, materials, processes, products) up to user applications, ensuring a consistent structure across implementations. The standard's reference architecture includes an entity-based data model and a set of functional building blocks for DT systems, adapted from general IoT architecture but specialized for manufacturing needs. This structured approach is significant because it offers manufacturers a unified blueprint: by aligning with ISO 23247, organizations can develop DTs with interoperability and scalability in mind, rather than using ad-hoc or proprietary architectures.

In current practice, ISO 23247 is increasingly used as a guiding framework in research and pilot applications of manufacturing DTs. For example, Kim et al. [54] implemented a DT architecture for a WAAM process based on ISO 23247, demonstrating how the standard's layered model helps address integration of sensors, machine controllers, and analytics for real-time monitoring and anomaly detection. Likewise, industry projects have begun aligning their DT platforms with ISO 23247; one study [55] demonstrated a drone factory assembly line integrating IoT and DT-compliant software, enabling real-time control, VR-assisted operations, and scenario simulations. Other applications include automotive assembly lines [56], industrial metaverse [57], and modular production systems [58]. This adoption across diverse manufacturing scenarios underscores ISO 23247's role as an organizational backbone, helping teams and vendors achieve consistency, interoperability, and comprehensive operational context. A summary of ISO 23247's core architectural components, along with their functions, benefits, and limitations, is presented in Table 5.2 to provide a consolidated view of the standard's structure and practical implications.

5.5.2 Limitations and Practical Challenges

Despite its promise, the ISO 23247 standard faces several limitations in practical adoption across industrial settings. To reflect both architectural and governance considerations, these challenges can be grouped into two broad categories: (1) implementation and interoperability issues, and (2) security, privacy, and data-governance challenges.

5.5.2.1 Implementation and Interoperability Challenges

- **Early Adoption and Awareness:** ISO 23247 adoption remains limited, as many digital twin implementations were developed independently. Ferko et al. [59] noted that industry often emphasizes only parts of the standard, reflecting low awareness and reluctance to overhaul existing architectures.
- **Complexity and Implementation Effort:** The standard is broad and abstract, requiring significant interpretation and design effort. Implementing ISO 23247 fully-especially in SME settings-can be resource-intensive due to the need to map entities and data flows without supporting tools.

Table 5.2 ISO 23247 digital twin framework: key components, functions, benefits, and limitations

Architecture component	Key benefits	Known limitations
Observable Manufacturing Elements (OME)	Enables complete, contextual digital modeling of factory operations	Not formally modeled in ISO 23247; requires domain-specific standards (e.g., OPC UA, MTConnect) for representation
Device Communication Entity	Enables modular integration of sensors and actuators; promotes plug-and-play connectivity	Lacks mandated protocols or interfaces; implementers must handle integration and interoperability challenges
Digital Twin Entity	Centralizes twin logic and analytics; supports scalable, fit-for-purpose implementations	Abstract design; lacks guidance on version control, persistence, and lifecycle validation
User Entity	Facilitates enterprise integration and stakeholder access to twin insights	No standardized APIs or user interaction models; requires custom integration
Cross-System Entity	Enhances interoperability and security; supports twin integration across domains	High-level guidance only; lacks concrete tools or reference ontologies
Traditional (Non-standardized) DT *(for comparison)*	Faster to implement for narrow use-cases; high design flexibility	Poor scalability, reusability, and interoperability; difficult to maintain or integrate

- **Gaps in Specificity and Interoperability:** ISO 23247 avoids prescribing specific technologies or protocols, which offers flexibility but causes ambiguity. The lack of guidance on data exchange, storage, and model versioning can lead to inconsistent and fragmented implementations.
- **Limited Tooling and Templates:** Few platforms or SDKs directly support ISO 23247, forcing users to manually integrate its concepts. For example, Kim et al. [54] had to custom-build an entire WAAM architecture due to the absence of turnkey solutions.
- **Scope Gaps:** The standard lacks depth on digital twin lifecycle management and validation practices. Wallner et al. [60] highlights that while it covers product lifecycle aspects, it overlooks how twins themselves evolve or maintain fidelity over time.

5.5.2.2 Security, Privacy, and Data-Governance Challenges

- **Fragmented Security Guidance:** Although ISO 23247 designates *"security and assurance"* within the Cross-System Entity, the guidance is high-level [61]. Implementers must independently define concrete controls for encryption, authentication, and access management across heterogeneous vendor systems.

- **Expanded Attack Surface:** Distributed DT ecosystems connect sensors, controllers, cloud platforms, and AI services, increasing exposure to threats such as data exfiltration, model tampering, and supply-chain compromise.
- **Authentication and Authorization:** Robust, mutually authenticated links between Device Communication Entities and DT services are needed, complemented by fine-grained Role-Based Access Control (RBAC) and Attribute-Based Access Control (ABAC) for simulation models, telemetry topics, and APIs.
- **Data Ownership and Governance:** Cross-vendor DT deployments blur data boundaries. Clear policies must define ownership, residency, retention, and sharing of process data, test patterns, and virtual-lab imagery to prevent intellectual-property leakage.
- **Privacy Preservation:** When DT data involves operators or production workflows, anonymization and federated-learning techniques can protect sensitive information while maintaining analytic utility.
- **Operational Hardening:** DTs require continuous security posture management, secure-by-default protocol profiles, signed model packages, software bill of materials (SBOM), and tested incident-response procedures-to ensure resilience throughout the twin lifecycle.

In summary, while ISO 23247 provides an invaluable high-level framework, its effective deployment depends equally on robust security and governance practices that ensure trust, accountability, and safe data exchange across the digital-twin ecosystem.

5.5.3 Future Directions and Opportunities

Looking ahead, there are several opportunities for evolving ISO 23247 to better support real-world manufacturing needs and to ease its adoption:

5.5.3.1 Refinement of the Reference Architecture:

The standard could be expanded or revised to address the identified gaps in its current architecture. Continuous evolution is already recognized as necessary—for example, Latsou et al. [62] argue that the ISO 23247 framework should be refined to include specifications for data storage, model versioning, and other implementation details that are currently omitted. Future editions of the standard (or companion technical reports) could provide more concrete guidelines or modules for these aspects, thereby reducing ambiguity. Incorporating guidance on how to verify and validate DTs (perhaps by defining quality metrics or compliance checkpoints within the framework) would help ensure that implementations maintain fidelity over time.

5.5.3.2 Improved Interoperability and Integration Guidelines:

To tackle interoperability challenges, ISO 23247 could explicitly reference or incorporate other standards and technologies. One opportunity is to define standardized interfaces or data models for common manufacturing scenarios. For instance, aligning with data exchange standards (like ISO 21597 for linked data containers or AutomationML) and industrial communication protocols (OPC UA, MQTT, etc.) in an official capacity would give implementers a clearer path to achieving interoperability. Additionally, developing an ontology or reference data schema for the *"observable manufacturing elements"* would allow different DT systems (even from different vendors) to understand each other's core concepts. Such enhancements would make it easier to plug components together and realize the *"digital thread"* across the four layers. Kim et al. [54] demonstrated this need in an additive manufacturing context by using an open platform with socket communication to connect robots and applications, and they highlight that further integration is still a work in progress.

5.5.3.3 Development of Tools and Reference Implementations:

An important opportunity is for the community (standards bodies, researchers, and industry consortia) to create reference implementation blueprints, software libraries, and toolkits aligned with ISO 23247. These might include open-source middleware that implements the standard's functional entities or reference DT models for typical manufacturing equipment following the ISO 23247 data structures. If manufacturers can download a basic ISO 23247-compliant DT template (for, say, a CNC machine or a robotic cell [63]) and adapt it to their needs, the barrier to adoption would drop significantly. Similarly, certification programs or testbeds could be established to validate that vendor products (e.g., an IoT platform or a simulation software) conform to ISO 23247's interoperability requirements. Such ecosystem support would transform ISO 23247 from a paper specification into a practical foundation that companies can readily build upon.

5.5.3.4 Enhancing Support for Lifecycle and Evolution:

To ensure longevity and relevance, the standard should consider the entire lifecycle of digital twin systems. This means guiding how to manage updates to the twin as the physical system changes (new sensors, machine upgrades, process modifications) and how to decommission or replace twin components gracefully. One future direction is to introduce a formal notion of a *"digital twin lifecycle"* within ISO 23247, possibly as a new part of the series. This could define states of a DT and standard procedures for data handover between states. By doing so, ISO 23247 would help practitioners plan for the evolution of their DTs in parallel with the physical assets' lifecycle, thereby maintaining the twins' accuracy and usefulness over the years. Wallner et al.'s proposal of a lifecycle meta-layer to handle DT updates is an example of the kind of concept that could be standardized [60]. Embracing such ideas would empower the standard to not only facilitate initial implementation but also long-term sustainability of DT deployments.

Table 5.3 Conceptual mapping of ISO 23247 entities to RAMI 4.0 functional layers

ISO 23247 Entity/Layer	RAMI 4.0 layer
Observable Manufacturing Elements (OME)	Asset layer
Device Communication Entity	Integration layer
Cross-System Entity *(Data Translation, Security, Assurance)*	Communication layer
Digital Twin Entity *(Simulation, Analytics, Control)*	Functional layers
User Entity *(Visualization, Decision Support)*	Business layer
Digital Twin Lifecycle Management *(Future ISO 23247 Extension)*	Lifecycle & value stream axis

5.5.4 Alignment with RAMI 4.0

While ISO 23247 provides a domain-specific framework for manufacturing DTs, it can be conceptually aligned with RAMI 4.0 (Reference Architecture Model for Industry 4.0), which defines a broader multi-layer structure for cyber-physical systems [64, 65]. Table 5.3 illustrates how ISO 23247 entities correspond to the hierarchical layers defined in RAMI 4.0, highlighting their complementary roles in interoperability and lifecycle management.

This alignment highlights that ISO 23247 focuses primarily on operational interoperability within the factory domain, while RAMI 4.0 introduces a lifecycle and organizational dimension extending from design to decommissioning. A combined perspective enables manufacturers to design digital-twin solutions that are both standards-compliant and lifecycle-aware, bridging shop-floor data exchange (ISO 23247) with enterprise-level business processes (RAMI 4.0). In parallel, the NIST Smart Manufacturing Framework (SMF) can also offer a complementary perspective by defining functional domains (e.g., connectivity, analytics, and decision support) that operationalize these architectural layers [66]. Future revisions of ISO 23247 may benefit from formally referencing RAMI 4.0 and harmonizing with frameworks like NIST SMF to achieve greater international coherence in digital-twin architectures.

5.6 Case Studies and Practical Implementations: Semiconductor Manufacturing

The semiconductor fabrication case study presented in this work is a conceptual example illustrating how multi-layer digital twins (DTs) can enable early yield prediction and process optimization.[2] Semiconductor fabrication, one of the most complex and capital-intensive

[2] This case study is a conceptual synthesis based on publicly available studies and the co-author's industrial experience in semiconductor process. No proprietary fab data were used; rather, representative process and AI-integration steps are outlined to enable methodological reproducibility.

manufacturing domains, is ideal for demonstrating the value of DTs that integrate sensor-rich process data, physics-based simulations, and AI/ML models. The following subsections highlight representative implementations and the potential of DTs in semiconductor manufacturing.

5.6.1 Multi-Layer Digital Twins for Yield and Process Control

Semiconductor manufacturing demands fast time-to-market, high-volume output, and high yield to remain profitable. Traditionally, fabs rely on statistical process control (SPC) and advanced process control (APC) techniques to monitor and adjust fabrication steps [67]. These methods use real-time data and statistical models to keep processes within spec, but they operate with limited scope (often per tool or step). A DT can greatly enhance this by providing an integrated, physics-rich simulation of the entire process flow, synchronized with sensor data across all tools. In fact, a DT can act as a replacement for conventional standalone simulations in process development, offering a continuously updated "virtual fab" that mirrors the real fabrication line. By aligning with the emerging ISO 23247 architecture, semiconductor fabs can develop these twins on a robust, interoperable blueprint.

In a semiconductor fabrication plant (fab), DT implementations can be envisioned in multiple layers or scales, each providing insights at different granularities. For example, at the process-tool level, a fab can maintain a "chamber-level" twin for equipment like etchers or deposition tools (including detailed physics models of plasma or chemical reactions), and a feature-level twin that simulates the patterning results on wafers. Building on these, a fab-wide twin integrates data from all tools and process stages. We describe this stratification as:

1. Discrete process simulation for predicting the yield of new process recipes from test wafers.
2. Cluster tool monitoring to calibrate and compare performance across tools from different vendors.
3. A comprehensive virtual fab that emulates each process step in sequence.
4. Continuous integration of module and process performance data to optimize yield in real time.

In practice, this means each tool (e.g., each lithography scanner or etcher) could have a localized twin that monitors its specific parameters and drifts. Such a twin can run virtual "experiments" for that tool, allowing engineers to fine-tune settings to minimize variation. At the next level, the comprehensive fab twin aggregates live data from all these tools and process steps-from wafer start to final sort-enabling prediction of final yield before the lot has even completed fabrication. This early visibility is crucial: if the DT forecasts a yield drop or systematic defect before final testing, engineers can intervene sooner (reworking or

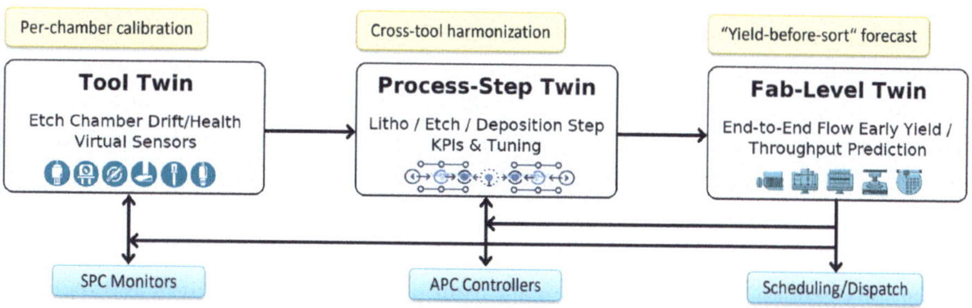

Fig. 5.8 Multi-layer twins and closed-loop yield control

adjusting later steps) to avoid scrapping large batches of wafers. Crucially, the integration of module-level and end-of-line data in a twin forms a feedback loop to continually improve the process *"recipe"* for high-volume manufacturing. By the time real wafers reach end-of-line tests, the twin (fed with in-line metrology and equipment data) has already projected potential failure modes, allowing engineers to adjust process parameters on the fly. Figure 5.8 shows an illustration of multi-layer DTs and closed-loop yield control.

5.6.2 AI-Augmented Fault Prediction and Defect Classification

After fabrication, semiconductor wafers undergo rigorous post-silicon testing and failure analysis to ensure only good chips ship to customers. This typically includes:

- **In-line metrology inspections:** performed during and after key process steps to detect structural or dimensional deviations;
- **Wafer-level electrical testing (E-test):** used to evaluate electrical characteristics and isolate failing dies;
- **Detailed failure analysis:** applied to suspect dies using optical, e-beam, or nanoprobing techniques.

Identifying the root cause of defects is incredibly challenging, given modern chips pack billions of transistors across dozens of layers [68]. Here, AI/ML models integrated into a DT can radically accelerate fault analysis and even predict issues before the final test. By embedding machine learning into the twins' analytical toolkit, the system can correlate patterns from manufacturing data and testing results to probable defect mechanisms. For instance, a DT that monitors comprehensive process parameters layer by layer can retain a *"signature"* of each wafer's journey; if a die fails at end-of-line, the twin can trace back through its simulated history to pinpoint which process step or tool setting likely introduced the flaw. Moreover, a *"virtual lab"* twin that aggregates automated test equipment (ATE) data, scan test patterns (ATPG results), and high-resolution images can use AI to predict the defect mode (e.g., electrical open, particle contamination, lithography error) before physical failure analysis is even performed. A recent study demonstrated a real-to-sim automatic simulation model generator for a DT in semiconductor manufacturing, validated across two operational fabrication facilities. The resulting DT simulations closely matched actual factory key performance indicators, with throughput and uptime deviations within 7.5% across machine groups during 300-day validation runs [69].

In parallel, inline defect classification during fabrication is being supercharged by AI. Modern fabs generate immense image data from optical wafer inspectors and metrology tools. Machine learning can learn the *"normal"* pattern of a wafer and automatically flag anomalies in real time. Unsupervised anomaly detection algorithms, for example, build a profile of what a defect-free wafer should look like and then identify any deviations without needing explicit training on every defect type. Embedding these capabilities in a DT means the twin is not just passively receiving data-it is actively interpreting it. In practical terms, an AI-infused twin can stop a lot mid-flow if it predicts a catastrophic defect is occurring, or route wafers to rework based on defect patterns.

Another powerful synergy is using high-fidelity simulations to generate synthetic data for ML-essentially the twin helping to train AI. Many fatal defect modes in chips are rare and not represented well in historical data, which limits supervised learning. A DT can fill that gap by simulating *"what-if"* scenarios or rare failure conditions (e.g., lithography misalignment, brief equipment glitches), producing virtual defect patterns. These simulated defects can be added to the ML training set, greatly improving the model's ability to recognize and diagnose them in real production [70]. By augmenting real data with twin-generated data, manufacturers gain a more robust defect prediction model that is prepared for even novel failure mechanisms. The practical examples of these approaches are summarized in Table 5.4.

5.6.3 Closed-Loop Feedback and Recipe Optimization

Digital twins in semiconductor manufacturing are progressing from simple monitoring tools into autonomous control systems. The vision is a fully optimized, closed-loop fab where the twin not only forecasts outcomes but also adjusts recipes and tool parameters in real time, effectively self-optimizing production. For example, a twin monitoring an etch process can detect drift in etch rate via virtual sensors and immediately recalibrate gas flow or RF power on the next wafer-without human intervention. This closed-loop approach extends traditional APC by combining AI/ML-driven predictions with physics-based simulation. Platforms such as Siemens MindSphere (linked with Siemens NX), Microsoft Azure Digital Twins, and plasma-etch twins have shown how IoT telemetry and calibrated models can reduce costly trial-and-error cycles [72]. Instead of repeatedly fabricating test wafers, adjustments are simulated virtually, providing faster-than-real-time recipe feedback and accelerating process qualification.

Table 5.4 Practical examples of AI/DT integration for defect detection and prediction

Use Case	Approach/Technology	Example/Source
Inline defect classification	Vision-based classification of wafer inspection images with anomaly detection	TSMC deployed advanced AI vision models within its DT/AI toolkit, accelerating automated wafer defect classification and root-cause pinpointing [71]
Lot-level anomaly detection	ML models learn "normal" wafer signatures and flag deviations in real time	Modern fabs integrate ML into DTs to stop lots mid-flow or reroute wafers predicted to have catastrophic defects
Synthetic defect generation for ML training	Physics-based DT simulations generate rare/what-if failure scenarios	Lithography or etch twins simulate misalignment or equipment glitches, producing synthetic defect data to improve ML robustness
Cross-domain demonstration	DT-simulation fused with real-time sensor data to predict failures in advance	Phoenix Contact engineers used *Ansys Twin Builder* with sensor fusion for predictive failure analysis in relays, showing transferability to fabs

A key enabler is cloud infrastructure, which allows supervisory twins to aggregate data across hundreds of tools and multiple fabs. Services like AWS IoT TwinMaker or Azure enable cross-fab comparisons, flagging out-of-spec behavior, and harmonizing recipes across equipment fleets. Parle et al. [16] highlighted how cloud-native scalability supports not only operational uniformity but also sustainability goals such as energy efficiency. Closed-loop twins also implement feed-forward control: issues detected early (e.g., lithography focus errors) are used to fine-tune downstream steps for specific wafers. This wafer-level personalization is impossible to do manually but achievable with algorithmic twins, leading to significant yield and throughput gains. As Moyne et al. [73] noted, run-to-run optimization twins can even be coupled with scheduling twins to optimize lot routing across the fab.

5.7 Conclusion

Digital twins are moving from isolated pilots to the backbone of smart manufacturing. This chapter has mapped that progression across three layers: (i) a tools landscape spanning cloud IIoT platforms, high-fidelity simulation, and PLM/virtual-twin suites; (ii) the ISO 23247 framework that enables compatible and scalable system implementations; and (iii) semi-conductor case studies that demonstrate multi-layer twins for yield prediction, AI-assisted defect classification, and closed-loop process optimization. Together, these threads show that the combination of continuous telemetry, physics-based models, and AI/ML transforms DTs from dashboards into predictive, prescriptive systems capable of real-time decision support and recipe tuning on the fab floor. At the same time, practical gaps remain. Industrial teams still face interoperability friction, limited out-of-the-box support for ISO 23247 entities, and uneven lifecycle governance of twin fidelity. Data quality, labeling effort for AI, and the scarcity of rare-defect examples are persistent barriers-although hybrid twins and synthetic data help close those gaps.

Looking forward, we see three priorities. First, deeper standardization-including reference models, testbeds, and conformance tooling for ISO 23247-will reduce time-to-value and improve portability. Second, higher autonomy via closed-loop control (from tool-level up to fab-wide scheduling) will push DTs toward self-optimizing operations. Third, human-centered interfaces, including LLM-enabled copilots, will make complex diagnostics and "*what-if*" exploration accessible to cross-functional teams. With these advances, digital twins will not only mirror factory reality but also continuously improve it, delivering resilient, energy-efficient, and economically compelling manufacturing at scale. By closing the loop, DTs convert manufacturing into a self-correcting system. They bring to life the Industry 4.0 vision of a smart fab that not only detects problems and diagnoses them (diagnostic and predictive analytics) but also prescribes and implements solutions autonomously. Achieving this at scale will require continued adherence to standards (to ensure all these moving parts interoperably share data) and robust validation, but the payoff is transformative: near-zero excursions, adaptive processes, and maximal yield and throughput.

References

1. Lasi, H. et al. 2014. Industry 4.0. *Business & Information Systems Engineering* 6(4): 239–242.
2. Thoben, K.-D., Stefan Wiesner, and Thorsten Wuest. 2017. Industrie 4.0 and smart manufacturing-a review of research issues and application examples. *International Journal of Automation Technology* 11(1): 4–16.
3. Tao, F. et al. 2019. Digital twin-driven product design framework. *International Journal of Production Research* 57(12): 3935–3953.
4. Kabir, Md Rafiul, and Sandip Ray. 2023. Virtual prototyping for modern internet-of-things applications: A survey. *IEEE Access* 11: 31384–31398.
5. Qi, Qinglin, et al. 2018. Digital twin service towards smart manufacturing. *Procedia Cirp* 72: 237–242.
6. Rasheed, Adil, Omer San, and Trond Kvamsdal. 2020. Digital twin: Values, challenges and enablers from a modeling perspective. *IEEE Access* 8: 21980–22012.
7. Liu, Mengnan, et al. 2021. A review of digital twin and its application in smart manufacturing. *IEEE Access* 9: 41192–41212.
8. Tao, F. et al. 2018. Digital twins and cyber–physical systems toward smart manufacturing and Industry 4.0: Correlation and comparison. *Engineering* 4(5): 653–661.
9. Zheng, Yiming, Shenhao Yang, and Haobo Cheng. 2021. Applications of artificial intelligence in digital twin: Review and prospects. *IEEE Access* 9: 156382–156398.
10. Bommasani, R. et al. 2021. On the opportunities and risks of foundation models. arXiv:2108.07258.

11. Kabir, Md Rafiul, and Sandip Ray. 2025. Digital twin tools for smart manufacturing: A paradigm shift for industry 4.0. *IEEE Open Journal of the Industrial Electronics Society* 6: 1756–1770.
12. Kurz, H.D. et al. 2022. *The Routledge Handbook of Smart Technologies: An Economic and Social Perspective*. Routledge.
13. Munirathinam. S. 2020. Industry 4.0: Industrial internet of things (IIOT). *Advances in Computers*, 117(1): 129–164. Elsevier.
14. Zion Market Research. 2024. *Digital Twins in Manufacturing Market Size, Share and Trends Analysis Report*. Accessed: 2025-04-15. https://www.zionmarketresearch.com/report/digital-twins-in-manufacturing-market.
15. Microsoft Azure. 2024. *Azure Digital Twins – Internet of Things (IoT) Platform*. Accessed: 2025-04-19. https://azure.microsoft.com/en-us/products/digital-twins/.
16. Parle, Dattatraya, et al. 2024. A comparative analysis for harnessing digital twin platforms for net-zero manufacturing. *IEEE Access* 12: 180175–180197.
17. Fahim, Muhammad, et al. 2022. Machine learning-based digital twin for predictive modeling in wind turbines. *IEEE Access* 10: 17976–17985. https://doi.org/10.1109/ACCESS.2022.3147602.
18. Lu, Q. et al. 2020. Digital twin-enabled anomaly detection for built asset monitoring in operation and maintenance. *Automation in Construction* 118: 103277.
19. Picone, M., Marco Mamei, and Franco Zambonelli. 2023. A flexible and modular architecture for edge digital twin: implementation and evaluation. *ACM Transactions on Internet of Things* 4(1): 1–27. https://doi.org/10.1145/3573206.
20. Lehner, D. et al. 2025. Model-driven engineering for digital twins: A systematic mapping study. In *Software and Systems Modeling*. https://doi.org/10.1007/s10270-025-01264-7.
21. Petrik, D., and Georg Herzwurm. 2019. iIoT ecosystem development through boundary resources: a Siemens MindSphere case study. In *Proceedings of the 2nd ACM SIGSOFT International Workshop on Software-Intensive Business: Start-Ups, Platforms, and Ecosystems*, 1–6.
22. Amazon Web Services. 2025. *AWS IoT TwinMaker*. https://aws.amazon.com/iot-twinmaker/. Accessed: 2025-04-30.
23. PTC Inc. 2025. *ThingWorx: Industrial IoT Software*. https://www.ptc.com/en/products/thingworx. Accessed: 2025-04-30.
24. Cline, B. et al. 2024. An IoT platform delivers expectations for a standardized PHM system framework. In *2024 IEEE International Conference on Prognostics and Health Management (ICPHM)*, 239–242. IEEE.
25. Doshi, S., Prajwal Ghatage, and Rahul Borole. 2023. IIoT for factory optimization and logistics management using PTC thingworx and kepware. In *2023 11th International Conference on Emerging Trends in Engineering & Technology-Signal and Information Processing (ICETET-SIP)*, 1–5. IEEE.
26. Bellalouna, Fahmi. 2021. Case study for design optimization using the digital twin approach. *Procedia CIRP* 100: 595–600.
27. Ansys Inc. 2025. *Ansys Twin Builder*. https://www.ansys.com/products/digital-twin/ansys-twin-builder. Accessed: 2025-04-21.
28. Pliuhin, V., Y. Tsegelnyk, and S. Plankovskyy. 2023. Implementation of induction motor speed and torque control system with reduced order model in ANSYS twin builder. In *International Conference on Reliable Systems Engineering*. Springer.
29. Pliuhin, V., V. Herasymenko, and A. Trotsai. 2023. Permanent magnet synchronous generator stabilization system with induction motor in ANSYS twin builder. In *International Conference on Smart Systems and Technologies*. Springer.
30. Tietieriev, V., V. Pliuhin, and V. Okhrimenko. 2023. Wind turbine permanent magnet generator speed stabilization system in ANSYS twin builder. In *International Conference on Smart Systems and Technologies*. Springer.

31. Vernova, GE. 2025. *Predix Overview*. https://www.gevernova.com/software/documentation/predix-platforms/PDFs/Predix_Overview.pdf. Accessed: 2025-04-30.
32. Dassault Systèmes. 2025. *The 3DEXPERIENCE Platform*. https://www.3ds.com/3dexperience. Accessed: 2025-04-30.
33. Bentley Systems. 2025. *iTwin Platform*. https://www.bentley.com/software/itwin-platform/. Accessed: 2025-04-30.
34. Autodesk Inc. 2025. *Autodesk Tandem*. https://intandem.autodesk.com/. Accessed: 2025-04-30.
35. Elias, C., Isabelle Southern, and Raja RA Issa. 2024. Digital twins for smart decision making in facilities and asset management (FAM). In *International Conference on Computing in Civil and Building Engineering*, 272–284. Springer.
36. Grunewald, J. 2022. *INVISTA uses AWS IoT TwinMaker for Connected Worker Operations*. AWS IoT TwinMaker Customers Page. https://aws.amazon.com/iot-twinmaker/customers/.
37. G. T. (quoted) Ju and N. (quoted) Cumins. 2023. Shell selects Bentley's iTwin platform to streamline capital projects. *J. Petrol. Technology (Data Science & Digital Engineering)*. https://jpt.spe.org/shell-taps-bentleys-itwin-platform-speed-project-delivery.
38. Kombaya Touckia, J., Nadia Hamani, and Lyes Kermad. 2022. Digital twin framework for reconfigurable manufacturing systems (RMSs): Design and simulation. *The International Journal of Advanced Manufacturing Technology*, 120(7): 5431–5450.
39. Cimino, C., Elisa Negri, and Luca Fumagalli. 2019. Review of digital twin applications in manufacturing. *Computers in Industry*, 113: 103130.
40. Mattera, G., Alessandra Caggiano, and Luigi Nele. 2025. Optimal data-driven control of manufacturing processes using reinforcement learning: an application to wire arc additive manufacturing. *Journal of Intelligent Manufacturing*, 36(2): 1291–1310.
41. Manta-Costa, A. et al. 2024. Machine learning applications in manufacturing-challenges, trends, and future directions. *IEEE Open Journal of the Industrial Electronics Society*.
42. Abadi, M. et al. 2016. TensorFlow: A system for Large-Scale machine learning. In *12th USENIX Symposium on Operating Systems Design and Implementation (OSDI 16)*, 265–283.
43. Imambi, S., Kolla Bhanu Prakash, and GR Kanagachidambaresan. 2021. PyTorch. In *Programming with TensorFlow: Solution for Edge Computing Applications*, 87–104.
44. Kramer, O., and Oliver Kramer. 2016. Scikit-learn. In *Machine Learning for Evolution Strategies*, 45–53.
45. Wang, P. et al. 2020. Digital twin-driven smart manufacturing: Connotation, reference model, applications and research issues. *Robotics and Computer-Integrated Manufacturing*, 61: 101837.
46. Tao, F. et al. 2022. Digital twin in industry: State-of-the-art. *IEEE Transactions on Industrial Informatics*, 18(6): 3567–3577.
47. Perno, M., Lars Hvam, and Anders Haug. 2023. A machine learning digital twin approach for critical process parameter prediction in a catalyst manufacturing line. *Computers in Industry*, 151: 103987.
48. Jeet Kaur, M., Ved P Mishra, and Piyush Maheshwari. 2020. The convergence of digital twin, IoT, and machine learning: Transforming data into action. In *Digital Twin Technologies and Smart Cities* (2020), 3–17.
49. Rathore, M. M. et al. 2021. The role of AI, machine learning, and big data in digital twinning: A systematic literature review, challenges, and opportunities. *IEEE Access* 9: 32030–32052.
50. Jin, L. et al. 2024. Big data, machine learning, and digital twin assisted additive manufacturing: A review. *Materials & Design*, 113086.
51. Jyeniskhan, Nursultan, et al. 2023. Integrating machine learning model and digital twin system for additive manufacturing. *IEEE Access* 11: 71113–71126.
52. Liu, Zheng, et al. 2023. The role of large language models in intelligent manufacturing. *Journal of Manufacturing Systems* 69: 523–536.

53. EOT. 2024. *Industrial Digital Twin – Transforming Industry with Real-Time Insights*. Accessed: 2025-04-19. https://eot.ai/industrial-digital-twin/.
54. Kim, D. B., Guodong Shao, and Guejong Jo. 2022. A digital twin implementation architecture for wire+ arc additive manufacturing based on ISO 23247. *Manufacturing Letters*, 34: 1–5.
55. Erdal, L. 2024. Integrating dynamic digital twins: Enabling real-time connectivity for IoT and virtual reality. In *Winter Simulation Conference (WSC)*, 2987–2998. IEEE.
56. Melo, V. et al. 2024. Design of an ISO 23247 compliant digital twin for an automotive assembly line. In *2024 IEEE 7th International Conference on Industrial Cyber-Physical Systems (ICPS)*, 1–6. IEEE.
57. Júnior, A.O. et al. 2025. Industrial metaverse digital twin: ISO 23247 compliant architecture for AI-driven simulation. In *2025 IEEE 8th International Conference on Industrial Cyber-Physical Systems (ICPS)*, 1–6. IEEE.
58. Le, G-N. et al. 2024. Multi-services digital twin for modular production system based on ISO 23247 and web server. In *International Conference on Sustainability and Emerging Technologies for Smart Manufacturing*, 679–689. Springer.
59. Ferko, E. et al. 2023. Standardisation in digital twin architectures in manufacturing' In *2023 IEEE 20th International Conference on Software Architecture (ICSA)*, 70–81. IEEE.
60. Wallner, Bernhard, et al. 2023. Digital twin development and operation of a flexible manufacturing cell using ISO 23247. *Procedia CIRP* 120: 1149–1154.
61. Belfadel, A., Stephen Creff, and Amira Ben Hamida. 2024. Advancing industrial digital twins: towards an open platform aligned with standards. In *IFIP International Conference on Product Lifecycle Management*, 110–124. Springer.
62. Latso, C et al. 2024. A unified framework for digital twin development in manufacturing. *Advanced Engineering Informatics* 62: 102567.
63. Vítor Arantes Cabral, J., Alberto José Álvares, and Guilherme Caribé de Carvalho. 2024. Digital twin implementation for an additive manufacturing robotic cell based on the iso 23247 standard. *IEEE Latin America Transactions* 22(8): 651–658.
64. Gottschalk, M., Mathias Uslar, and Christina Delfs. 2017. *The use case and smart grid architecture model approach: The IEC 62559-2 use case template and the SGAM applied in various domains*. Springer.
65. Grangel-González, T. et al. 2016. An RDF-based approach for implementing industry 4.0 components with Administration Shells. In *2016 IEEE 21st International Conference on Emerging Technologies and Factory Automation (ETFA)*, 1–8. IEEE.
66. Shao, Guodong, and Moneer Helu. 2020. Framework for a digital twin in manufacturing: Scope and requirements. *Manufacturing Letters* 24: 105–107.
67. Ghelani, H. 2024. Advanced AI technologies for defect prevention and yield optimization in PCB manufacturing. *International Journal Of Engineering And Computer Science* 13(10).
68. Liebens, M. et al. 2018. In-line metrology for characterization and control of extreme wafer thinning of bonded wafers. *IEEE Transactions on Semiconductor Manufacturing* 32(1): 54–61.
69. Behrendt, S. et al. 2025. Real-to-sim: automatic simulation model generation for a digital twin in semiconductor manufacturing. *Journal of Intelligent Manufacturing*, 1–20.
70. Sengodan, T., Sanjay Misra, et al. 2025. *Advances in Electrical and Computer Technologies: Proceedings of the 6th International Conference on Advances in Electrical and Computer Technologies (ICAECT 2024), Tiruchengode, India, September 26th–27th, 2024*. CRC Press.
71. Taiwan Semiconductor Manufacturing Company (TSMC). 2025. *IntelligentFab® Automation – Intelligent Packaging Fab*. Accessed: 2025-08-24. https://www.tsmc.com/english/dedicatedFoundry/services/apm_intelligent_packaging_fab/intelligentFab_automation.

72. Lau, C.L. et al. 2025. Process-aware digital twins by deep learning for DUV photolithography and plasma etch. *IEEE Transactions on Semiconductor Manufacturing*.
73. Moyne, J. et al. 2020. A requirements driven digital twin framework: Specification and opportunities. *IEEE Access* 8.

The manufacturer's authorised representative in the EU is Springer Nature Customer Service Centre GmbH, Europaplatz 3, 69115 Heidelberg, Germany. If you have any concerns regarding our products, please contact ProductSafety@springernature.com

Printed and bound by CPI Group (UK) Ltd, Croydon, CR0 4YY
07/07/2026
02160926-0006